ROOTS TO SEEDS

ROOTS TO SEEDS

400 YEARS OF OXFORD BOTANY

Stephen A. Harris

Bodleian Library
UNIVERSITY OF OXFORD

CONTENTS

FOREWORD

Founded 400 years ago, Oxford Botanic Garden is the birthplace of botanical science in Oxford. It is not surprising, then, that over those four centuries Oxford has also built up an outstanding collection of plant specimens, botanical illustration and rare books on plant classification, collecting and plant biology.

These collections, and the living plants in the Garden, have been integral to the study of botany in the university. The Physic Garden, as it was then known, was founded in 1621 to support medical training at Oxford. In 1648, the first catalogue of the Garden was published, which recorded nearly 1,400 different plants, neatly demonstrating the connections between books, the study of botany and the dissemination of knowledge about plants.

The exhibition which this book accompanies explores the diverse heritage of botanical sciences in the university, including the rich manuscript and rare book collections at the Sherardian Library of Plant Taxonomy, the Botanic Garden and Arboretum, the Oxford Herbaria and the Bodleian Libraries.

These include some of the late eighteenth-century sketches of plants which the celebrated botanical artist, Ferdinand Bauer, brought back with him from his exploration of the eastern Mediterranean with John Sibthorp, Sherardian Professor of Botany. In Oxford, from these sketches, Bauer drew 966 superbly hand-coloured illustrations for Sibthorp and Smith's *Flora Graeca*, which the renowned botanist Joseph Hooker called 'the greatest botanical work that has ever appeared'.

Seventeenth-century herbals and elegant garden plans, fossil slides, plant models, beautifully preserved specimens and early manuscript attempts at plant classification all help to tell the story of the development of botanical science in Oxford and some of the intrepid botanists who devoted themselves to the essential study of plants. This is a living history which continues, encompassing pioneering botanical research and a plant collection which thrives in the heart of Oxford.

I would like to thank curator and author Stephen Harris for all he has done to make this exhibition possible, together with colleagues at the Botanic Garden and Arboretum, the Department of Plant Sciences, the History of Science Museum and the Bodleian Library exhibitions team, particularly Madeline Slaven and Sallyanne Gilchrist. I am also very grateful to the Friends of Oxford Botanic Garden and the Danby Patrons' Group for their generous support.

Simon Hiscock
Director, Oxford Botanic Garden and Arboretum

PREFACE

In nature, a plant's survival is a consequence of its genetic make-up and the interactions of those genes with the environment. When environments change, previously successful plants may find themselves struggling for survival, unless they have suitable resilience. This may be a metaphor for botany at the University of Oxford, where genetic make-up is the staff and the students, who generate botanical ideas and enthusiasm, and the environment is determined by leadership, university engagement, and national and international communities.

Marking the foundation of the Oxford Botanic Garden in 1621 is an opportunity to reflect on the contribution that four centuries of botanical research in the university has made to our understanding of plant biology. This book, and the exhibition that accompanies it, reveal the collaborative and competitive aspects of botany, and the central positions professors of botany have had in shaping botanical legacies. Consequently, botanical activity across the history of the university is patchy; long periods of relative inactivity are punctuated by bursts of intensely productive research. The archives are often replete with the names and activities of professors, and silent regarding the hundreds of people who supported their work directly or indirectly, whether they were specimen collectors in the field (including, under eighteenth-century colonialism, slaves), horticulturalists in the garden or technicians in the laboratory. Botany at Oxford is unexceptional in this regard, though current projects rightly strive to encompass the people who have traditionally remained unacknowledged.

Rather than providing a chronological sequence of events, this selective account, which makes no claims to be comprehensive, is divided into seven chapters focused on specific themes. Chapter 1 introduces the context and early history of botany at Oxford. Chapter 2 focuses on the university's botanical collections. Chapter 3 considers how objects in the collections arrived in the university over the last four centuries. Chapters 4 and 5 are about the contributions Oxford has made to botanical research, nationally and internationally, arbitrarily divided between morphology and taxonomy on the one hand, and broadly experimental approaches on the other. In Chapter 6 the focus is on applied botany and the problems faced by agriculture and forestry in the university. The final chapter is concerned with four centuries of botanical teaching at Oxford.

During the last fifty years, Oxford-based botanists have contributed knowledge across a great diversity of areas within botany and, more broadly, the plant sciences. However, time is needed for the botanical community to rigorously assess individual contributions. Therefore, rather than attempt to review botanical research and teaching conducted at Oxford comprehensively, particularly over the last fifty years, I have presented snapshots reflecting the types of contributions that Oxford-based botanists have made to the global collaboration that is modern plant sciences.

If the last 400 years have shown anything, it is that botany at Oxford succeeded by focusing on quality research and by equipping subsequent generations with the tools necessary to frame and answer relevant questions.

AUTHOR'S NOTE

Until 1834, the Oxford Botanic Garden was called the Physick or Physic Garden. For clarity, I have used 'Botanic Garden' or 'Garden' irrespective of the period under consideration. After 1963 the term 'Botanic Garden' or 'Garden' is understood to include the Arboretum. The term 'Department of Plant Sciences' refers to the department created when the Departments of Botany, Forestry and Agricultural Sciences were merged in 1985. Quotations are as in original texts, except that where necessary the long ∫ has been converted to the modern s and ligatures have been split. Clarifications are added in square brackets. All prices are quoted from original documents.

Questions frequently arise about values, hence conversions to 2020 equivalents have been given for guidance.[1] To avoid unnecessary confusion, the person responsible for the practical work in the Garden was the 'superintendent', and until 2002 academic responsibility lay with the Sherardian professor – after this date both roles were taken by the director of the Garden. For a general outline of events associated with plant sciences in Oxford, see the Timeline.

I am grateful for access to manuscripts and photographs preserved in the Bodleian Library, the Botanic Garden and Arboretum, the British Library, Oxford University Herbaria, the Department of Plant Sciences and the Royal Society. It is my pleasure to thank many people for their direct and indirect contributions to this book: John Baker, Anne Marie Catterall, Liam Dolan, Samuel Fanous, Barrie Juniper, Serena Marner, Janet Phillips, Leanda Shrimpton, Sophie Wilcox, Rosemary Wise and the members of the Oxford University and Harcourt Arboretum Florilegium Society, and an anonymous reader of the manuscript. Of course, errors and misconceptions are mine, as are the views expressed.

TIMELINE

1621 Oxford Botanic Garden founded

1642 Jacob Bobart the elder (*c*.1599–1680) appointed as Garden superintendent

1648 First catalogue of the contents of the Garden published

1669 Robert Morison (1620–1683) appointed as Danby and Regius professor

1679 Jacob Bobart the younger (1641–1719) appointed as Garden superintendent

1720 Edwin Sandys appointed as professor of botany

1724 Gilbert Trowe appointed as professor of botany

1734 Johann Jacob Dillenius (1687–1747) appointed as first Sherardian professor

1738 James Smith appointed as gardener

1747 Humphrey Sibthorp (*c*.1713–1797) appointed as second Sherardian professor (until 1784)

1750 Georg Dionysius Ehret (1708–1770) appointed as gardener

1752 William Stockford appointed as gardener

1753 Thomas Potts appointed as gardener

1756 John Foreman appointed as gardener; he relinquished the position *c*.1790

1784 John Sibthorp (1758–1796) appointed as third Sherardian professor

1793 John Sibthorp appointed as Regius professor

1796 George Williams (*c*.1762–1834) appointed as fourth Sherardian and Regius professor

1813 William Baxter (1787–1871) appointed as Garden superintendent

1834 Charles Giles Bridle Daubeny (1795–1867) appointed as fifth Sherardian professor

1840 Daubeny appointed as Sibthorpian professor

1851 William Hart Baxter (*c*.1816–1890) appointed as Garden superintendent

1853 Maxwell Tylden Masters (1833–1907) appointed as Fielding Curator (until *c*.1856)

1868 Marmaduke Alexander Lawson (1840–1896) appointed as sixth Sherardian professor and Sibthorpian professor

1884 Isaac Bayley Balfour (1853–1922) appointed as seventh Sherardian professor

Joseph Henry Gilbert (1817–1901) appointed as Sibthorpian professor (until 1890)

1886 Selmar Schönland (1860–1940) appointed as Fielding Curator (until *c*.1889)

1888 William George Baker (1861–1945) appointed as Garden superintendent

Sydney Howard Vines (1849–1934) appointed as eighth Sherardian professor

1894 Robert Warington (1838–1907) appointed as Sibthorpian professor (until 1897)

1895 George Claridge Druce (1850–1932) appointed as Honorary Fielding Curator

1905 Imperial Forest School transferred to Oxford from Cooper's Hill, Surrey, with William Schlich (1840–1925) as head

1906 William Somerville (1860–1932) appointed as Sibthorpian professor

1907 School of Rural Economy established

1920 Frederick William Keeble (1870–1952) appointed as ninth Sherardian professor

Robert Scott Troup (1874–1939) appointed as head of School of Forestry

1924 Imperial Forestry Institute established, with Robert Troup as founding director

1925 James Anderson Scott Watson (1889–1966) appointed as Sibthorpian professor

1927 Arthur George Tansley (1871–1955) appointed as tenth Sherardian professor

1937 Theodore George Bentley Osborn (1887–1973) appointed as eleventh Sherardian professor

1939 School of Forestry and Imperial Forestry Institute merged, under directorship of Harry George Champion (1891–1979)

Nicholas Vladimir Polunin (1909–1997) appointed as Fielding Curator (until 1947)

1940 John Frederick Gustav Chapple (1911–1990) appointed as Druce Curator (until 1947)

1942 George William Robinson (1898–1976) appointed as Garden superintendent

1945 Geoffrey Emett Blackman (1903–1980) appointed as Sibthorpian professor

School of Rural Economy renamed Department of Agriculture

1947 Imperial Forestry Institute renamed Commonwealth Forestry Institute

1948 Edmund Frederic Warburg (1908–1966) appointed as Druce Curator, then Druce and Fielding Curator in 1957

1952 Edmund André Charles Louis Eloi Schelpe (1924–1985) appointed as Fielding Curator

1953 Cyril Dean Darlington (1903–1981) appointed as twelfth Sherardian professor

Sheila Littleboy appointed as Fielding Curator (until 1956)

1956 University field station at Wytham established

1959 Malcolm Vyvyan Laurie (1901–1973) appointed as head of Commonwealth Forestry Institute

1961 Frank White (1927–1994) appointed as curator of Forestry Herbarium, then Druce and Fielding Curator in 1971

1963 Kenneth Burras appointed as Garden superintendent

1968 Adrian John Richards appointed as Druce Curator (1968–70)

1969 Jack Harley (1911–1990) appointed as head of Commonwealth Forestry Institute

1970 John Harrison Burnett (1922–2007) appointed as Sibthorpian professor

1971 Robert Whatley (1924–2000) appointed as thirteenth Sherardian professor

1980 David Smith (1930–2018) appointed as Sibthorpian professor (until 1987)

Duncan Poore (1925–2016) appointed as head of Commonwealth Forestry Institute

1982 Jeffery Burley appointed as head of Commonwealth Forestry Institute

1983 Commonwealth Forestry Institute renamed Oxford Forestry Institute

1985 Department of Plant Sciences formed by combining the Departments of Agriculture, Botany and Forestry

1988 Timothy Walker appointed as Garden superintendent

1990 Christopher John Leaver appointed as Sibthorpian professor

1991 Hugh Dickinson appointed as fourteenth Sherardian professor

1992 David Mabberley appointed as acting curator of Herbaria

1994 Quentin Cronk appointed as acting curator of Herbaria

1995 Stephen Harris appointed as Herbaria curator

2002 Timothy Walker appointed as Garden director (2002–13) Oxford Forestry Institute closed

2007 Nicolas Harbard appointed as Sibthorpian professor

2009 Liam Dolan appointed as fifteenth Sherardian professor (until 2020)

2011 Wood Professor of Forest Science endowed

2013 John Mackay appointed as Wood professor

2014 Alison Foster appointed as acting Garden director

2015 Simon Hiscock appointed as Garden director

1 ROOT

Origins

Science is a cultural activity that helps us to make sense of the natural world.[1] Stories told about plants, such as those that surround the mandrake, are not science unless they survive rigorous objective testing. Scientific ideas are also moderated by the societies in which they emerge, and by reactions to those ideas; technologies, philosophies and prejudices may constrain scientists. Our understanding of science is therefore limited by where, when and how we live. Even Ferdinand Bauer, the late eighteenth-century artist who prized botanical accuracy, was not immune to the whimsy of myth when depicting the mandrake.

Pre-sixteenth-century botany

Scientific approaches to the study of plants, frequently justified in terms of the essential roles they play in the lives of people, were shaped in ancient Greece and Rome.[2] Greek philosophers (for example, Theophrastus), theorized about the nature of plants, while Roman writers focused on the application of plants to medicine (for example, Dioscorides), and agriculture and horticulture (for example, Varro and Cato the elder). Theophrastus' *Enquiry into Plants* and *On the Causes of Plants*, written between about 350 BCE and 371 BCE, are collections of lecture notes discussing plant anatomy, physiology, morphology, ecology and classification.[3] In contrast, Dioscorides' *De materia medica* (50–70 CE) is a practical, illustrated work on medicinal plants which, for nearly 1,500 years, was the basis of our knowledge about the therapeutic properties of plants, as it was copied, translated, recopied and abstracted.[4]

During the first millennium of the Common Era, knowledge of the Greek language declined in western Europe.[5] In the early centuries of the second millennium CE, when Greeks, Muslims and Christians came into contact around the Mediterranean, Greek science was rediscovered.[6]

The Danby Gate, from Rudolph Ackermann's *History of Oxford* (1814). Private collection.

THE MANDRAKE

The mandrake[7] (*Mandragora officinarum*), a native of the Mediterranean, has a rosette of large, glossy, green, broadly spear-shaped leaves with purplish bell-shaped flowers in late winter. By early autumn, its flowers will have developed into yellow fruits with a sweetish, rather resinous scent, the shape and size of table-tennis balls. Below ground, the parsnip-like root sometimes has an anthropoid form, giving rise to the name 'mandrake'.

Mandrake is an effective medicinal plant that has been used in Europe and the Middle East for millennia. Low doses of alkaloid-rich extracts produce drowsiness and anaesthesia in humans, moderate doses induce hallucinations, while high doses can kill. The boundaries between dosages are often narrow. If mandrake is to be used safely, users must realize that individual plants produce different amounts of alkaloid depending on where they are grown and when they are harvested.

Few plants in Western botany are surrounded by as much folklore, mythology and sheer nonsense as the mandrake. Speculation, wishful thinking, hand-me-down tales, protectionism and ritual have imbued it with an awesome reputation. When the homunculus-like root was ripped from the ground, 'old wives', 'runnagate Surgeons' and 'physicke-mongers'[8] would have us believe, its scream would kill the person who uprooted it. Incantations with swords and magic circles might protect harvesters, although a better method for harvesting was by using a dog; pre-Renaissance mandrake illustrations often come with a canine companion.

As early as 300 BCE, the Greek philosopher Theophrastus heaped ridicule on such rigmaroles. However, mandrakes, especially those with a strong human form, became desirable talismans across Europe. Rarity, and the associated tales, created a market, which was supplied by entrepreneurs dealing in mock mandrakes. Museums are replete with such fakes, often whittled from the roots of common plants such as white bryony.[9]

The dangers of unearthing mandrakes notwithstanding, they were adornments in British gardens as early as the tenth century. Years of practical experience of digging mandrakes from his garden in Holborn taught the sixteenth-century English herbalist John Gerard there was no peril in uprooting them,[10] whatever popular or ancient authority might assert. In the mid-seventeenth century, mandrakes were among the first plants grown in the Oxford Botanic Garden,[11] as knowledge about plants began to rely on practical observation rather than received wisdom.

Watercolour of a mandrake (*Mandragora officinarum*) by Ferdinand Bauer, based on field sketches he made during his journey in the eastern Mediterranean with John Sibthorp, and completed in Oxford between 1788 and 1792. Bodleian Library, Sherardian Library of Plant Taxonomy, MS. Sherard 244, f.53.

The Schola Medica Salernitana, a medical school near the southern coastal city of Salerno, Italy, was particularly important. A community of scholars emerged that was interested in translating ancient Greek manuscripts on plants and many other subjects, held in the Arab world, into Latin.[12] Ancient Greek knowledge was rediscovered in western Europe, spurring humanist ideas of the Renaissance.[13]

Whereas ancient Greek philosophies viewed humans as part of nature, Christian doctrine saw humans as being apart from nature.[14] The earth was at the centre of a universe that had been fundamentally created for humans; everything was subservient to the caprice of *Homo sapiens*. Christian doctrine also held that the world had fallen into corruption, and regarded the natural world as a backdrop to humanity's spiritual quest.

With the flowering of interest in the natural world came the establishment of modern physic gardens (collections of medicinal plants) in Italian cities such as Pisa, Padua and Bologna in the 1540s, and in Leiden in the Netherlands in the 1570s.[15] In early sixteenth-century London, large gardens were also maintained by private individuals such as the English apothecary John Gerard and the physician John Parkinson.

Interest in plants at Oxford did not start with the planting and blessing of a rock in a field on 25 July 1621, when the Oxford Botanic Garden was formally established. Before the foundation of the Botanic Garden, with its remit to teach physicians about medicinal plants, fellows of the university's colleges were investigating plants. Plants were usually studied in the context of medicine, but they also had strong religious associations; nature revealed God's goodness and care for people.[16]

The best botanical work in this period was being done in Continental Europe, but, given plants' strong associations with medicine and theology, it is unsurprising that some early British botanists should have Oxford connections; for example, the sixteenth-century English physician William Turner, 'father of British botany', was made a doctor of medicine at Oxford, and the botanist Henry Lyte was educated at Oxford.[17] Academic knowledge was synthesized into herbals, where plants were named, described, classified and illustrated from nature.[18] Yet the botanical knowledge of the populace, engaged in the day-to-day routine of growing plants to sustain all aspects of European life before the Renaissance, is poorly known. What has been recorded is selective – what academic writers of the time considered worthy of preservation.

Title page of John Gerard's *The Herball, or, Generall Historie of Plantes* (1597), one of the standard reference works in English used by the Bobarts as they began stocking Oxford Botanic Garden. Bodleian Library, Sherardian Library of Plant Taxonomy, Sherard 649.

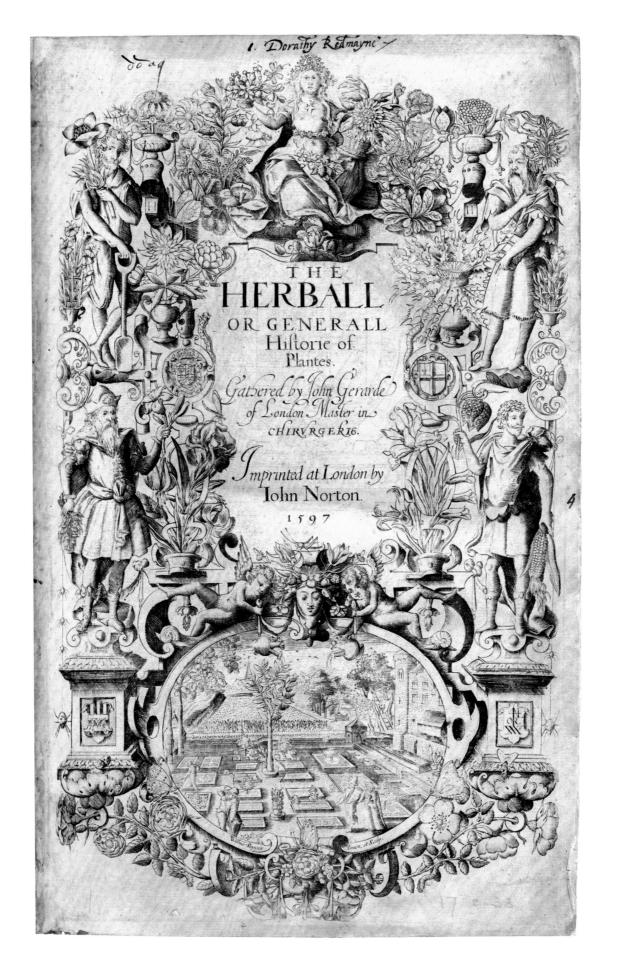

THE
HERBALL
OR GENERALL
Historie of
Plantes.

Gathered by John Gerarde
of London Master in
CHIRVRGERIE.

Imprinted at London by
Iohn Norton.
1597

Ways of thinking

From the fifteenth century onwards, new discoveries about the natural world challenged ancient authority. For example, the New World was populated by peoples and organisms that were not mentioned in ancient manuscripts or the Bible. Some people met such challenges by carefully finessing arguments, aligning evidence to fit the existing orthodoxy. Others, such as the English philosopher and statesman Francis Bacon, 'father of the scientific method', believed that natural philosophers must be methodical, sceptical and willing to abandon handed-down authority. Rather than taking an unproven hypothesis and reasoning from the general to the specific to test empirical observations, Bacon's scientific method of reasoning starts with specific observations to arrive at probable, general hypotheses.[19]

New ways of collecting evidence aid the scientific method, while the organization of like-minded individuals leads to ideas spreading and becoming established. In his unfinished utopian fantasy *The New Atlantis* (1627) Francis Bacon described Bensalem, a society characterized by 'generosity and enlightenment, dignity and splendour, piety and public spirit' and ruled by a wise, self-appointed elite on the basis of scientific principles. At this society's heart was Salomon's House, a pure and applied research organization equipped with a vast range of facilities including 'two very long and fair galleries'. One of these contained statues of inventors, the other 'patterns and samples of all manner of the more rare and excellent inventions'.[20] Salomon's House also included the ideal botanic garden:

> large and various orchards and gardens, wherein we do not so much respect beauty, as variety of ground and soil, proper for divers trees and herbs; and some very spacious, where trees and berries are set ... In these we practise likewise all conclusions of grafting, and inoculating as well of wild-trees as fruit-trees, which produceth many effects. And we make, by art ... trees and flowers to come earlier or later than their seasons, and to come up and bear more speedily than by their natural course they do; we make them also, by art, greater much than their nature, and their fruit greater and sweeter, and of differing taste, smell, colour, and figure from their nature; and many of them we so order that they become of medicinal use.[21]

The ideal of Salomon's House appeared to be fulfilled when the Royal Society was founded at Gresham College in London in 1660.[22] Experimentation was central to Bacon's 'new philosophy' and to the founders of the Royal Society, which included John Wilkins, warden of Wadham College, Oxford. Despite their belief in the importance of objective evidence for reaching conclusions about the world, members of the society, like the general populace, held a diversity of beliefs derived from religion, mythology and mysticism.[23]

The Royal Society was quick to recognize the importance of ensuring that its ideas were widely communicated. Within five years of its foundation, it produced its first major publication, Robert Hooke's *Micrographia, or, Some Physiological Descriptions of Minute Bodies Made by Magnifying Glasses, with observations and inquiries thereupon* (1665).[24] This scientific bestseller inspired people to use the microscope to explore nature, and to make inferences about how the world works. In 1668 the University of Oxford established a central printing workshop in the bowels of the Sheldonian Theatre and learned the power of print to control the flow of ideas and perhaps even make money.[25]

Supporters of established ideas naturally felt threatened by the new philosophy.[26] For example, in 1667 Robert South, a future canon of Christ Church, preached at Westminster Abbey that

> it cannot but be matter of just indignation to all knowing and good men, to see a company of lewd, shallow-brained huffs, making atheism and contempt of religion, the sole badge and character of wit, gallantry, and true discretion ... censuring the wisdom of all antiquity, scoffing at all piety, and, as it were, new modelling the whole world.[27]

Two years later, as the university's public orator, its official voice, he reinforced his views at the opening of the Sheldonian Theatre: 'it [South's sermon] was very long, & not without some malicious & undecent reflections on the *Royal Society* as underminers of the university, which was very foolish and untrue.'[28] His audience included influential members of the society, such as Elias Ashmole (the antiquary whose collection was the foundation for the Ashmolean Museum), John Wallis (a former warden of Wadham College and a founder member of the society), Christopher Wren (architect of the

Sheldonian and also a founder member) and John Evelyn (diarist, scholar and gardener).

By the mid-twentieth century, ideas of how science should be done shifted yet again. The value of the objective testing of hypotheses crystallized in the concept of 'falsifiability'. A theory about the world must produce hypotheses that are capable of being tested – and falsified. A theory is accepted until hypotheses arising from it are rejected. Today, scientists work in many different ways, but central to what they do is the precept that through an interactive process of observation, hypothesis, experimentation, evaluation and retesting the natural world can be understood.[29] Science is fluid, constantly evolving and changing our view of the world as new facts are accommodated within existing ideas or ideas that are unsupported by facts are jettisoned. For example, two fundamental biological ideas, Darwinian evolution and Mendelian genetics, emerged in the latter half of the nineteenth century.[30] As these ideas matured in the early twentieth century, they transformed how we think about living organisms, ourselves and our place in the world.[31]

As mentioned above, scientists do not operate in isolation, but are influenced by their predecessors, their peers and the places, times and societies in which they live.[32] Traditionally, knowledge of plants was separated between scholars, who thought about plants, and artisans (such as gardeners, foresters and farmers), who worked with plants. Modern scientists do not make such distinctions as they reveal ever more subtle details about life, through imaging, chemical or molecular technologies, or the power of computing and mathematics.

Global events such as political conflicts, the commercial demands of European empires, and discoveries about the medicinal and economic potential of plants such as quinine (*Cinchona* species) and rubber (*Hevea brasiliensis*) have been important drivers of botanical research.[33] However, local, internal events have done the most to shape the University of Oxford's contributions to botany over the past four centuries.

Professors in the university have made greater or lesser contributions to botany during their tenures.[34] Those who have occupied professorial chairs, and assumed the temporary responsibilities they bring with them for stewardship and leadership in botany, have been all manner of characters from the inept to the dynamic, from the autocrat to the democrat and from the collegiate to the self-interested.

Magnified surface of the leaf of a common nettle from Robert Hooke's *Micrographia, or, Some Physiological Descriptions of Minute Bodies Made by Magnifying Glasses, with observations and inquiries thereupon* (1665). Bodleian Library, Lister E 7, Schem XV.

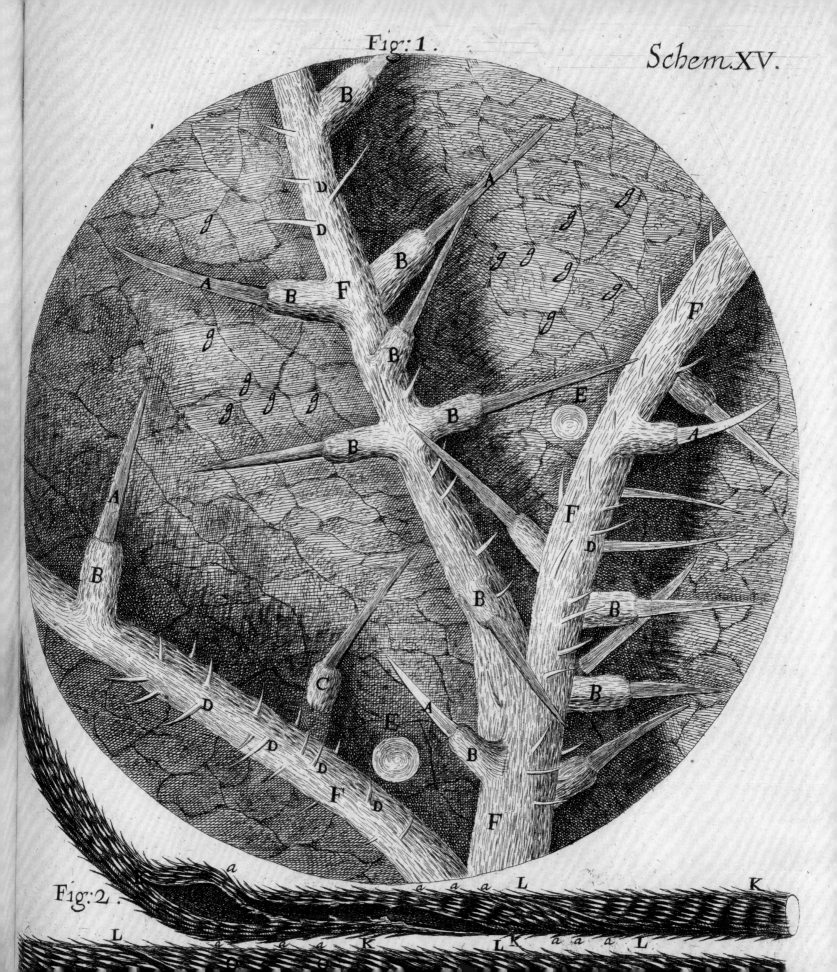

Fig: 2.

Intermittent botanical activity in the university has been attributed to apparent hostility by an 'arts university' to the sciences, the distance of Oxford from London (in the eighteenth century) and from centres of industrial innovation (in the nineteenth century), and its role as 'a finishing school for the sons of gentlemen'.[35] At least equally important was that some professors from outside Oxford, arriving with great ideas for change, were eventually cowed by the realities of the institution.

Foundation of the Oxford Botanic Garden

The Roman author Pliny complained about physicians' attitudes to learning about plants; it was 'more pleasant to sit in a lecture-room engaged in listening, than to go out into the wilds and search for the various plants'.[36] Medical training at Oxford had changed from its medieval origins of rote learning but there was still little practical instruction, other than the infrequent dissections of executed prisoners.[37] In 1620 'motions were made for the founding of a ... Garden for Physical Simples'[38] in the university to support the teaching of medicine. A botanic (physic) garden meant that medicinal plants would be within easy reach to help students, and their teachers, at least to recognize routinely prescribed plants.[39]

The following year, the wherewithal to establish the Garden was provided by Sir Henry Danvers, Earl of Danby. We do not know why Danvers wanted to support a botanic garden, but he may have been inspired by gardens he had seen during exile in France or simply desired to make a lasting mark. He donated £5,250 (c.£690,500 in 2020) to purchase, from Magdalen College, the lease of a five-acre (approximately two hectares), flood-prone field, build walls around it and begin the Oxford Botanic Garden.

At two o'clock, on the afternoon of Sunday, 25 July 1621, the university's vice chancellor, together with members of the university and its colleges, processed in their ceremonial finery from the Church of St Mary the Virgin along the High Street to Danvers's leased meadow. Following speeches by the university's senior members, a foundation stone was laid and some coins cast over it.[40]

With the ceremony complete, an unpromising site was gradually transformed into a garden fit for a 'City of Palaces'.[41] Danvers's money was spent on 'well fair and sufficient'[42] walls and gates to surround the plot. The elaborate Danby Gate, named to honour Henry Danvers, which was completed in 1632, cost £500 (c.£61,000 in 2020).[43] Features on the gate's north face, including a bust of Danvers, highlighted the Garden's patronage. The gate's inscription, 'Gloriae Dei Opt. Max. Honori Caroli Regis In Usum Acad. et Reipub. Henricus Comes Danby D.D. MDCXXXII',[44] emphasized the glory of God and Charles I, and the utility of the Garden for the university and the country, and immortalized the name of Danvers.

Between 1621 and 1636, as the masonry was being erected, the level of the site was raised above the adjacent river, with the help of Mr Windiat, who as the university's official scavenger was responsible for its waste disposal and provided '4000 loads of mucke & dunge'.[45]

A stage had been created upon which plant sciences would develop in the university, although there was no money to do it yet. Danvers's benefaction was exhausted by stone and mortar and 'mucke'.

A base for future science

The site of the Botanic Garden was finally separated from the city in 1633 when the wall was completed. It had taken a dozen years since the first stone was laid to get to this stage. By 1636 the gardener to Charles I, John Tradescant the elder, was confirmed as gardener in the Oxford Physic Garden, but he appears never to have taken up the post, or else his impact was minimal.[46] Only after Jacob Bobart the elder, an ex-soldier and publican, was appointed in 1642 are there records of plants appearing within the wall's precincts; we do not know how the land was used before this. However, whether the Garden could become what its founder had envisioned was in the balance. Funding was limited, as the settlement Danvers had made to finance it quickly proved inadequate.[47]

Moreover, there was civil war when Bobart the elder become head of the Botanic Garden, and Oxford was a garrison town. Charles I had established his court in Oxford following his retreat from London in the early stages of the First English Civil War (1642–46). During the Commonwealth, the university and colleges adapted, as they did again when Charles II was restored to the throne in 1660.

Several unrelated events in Italy and England in 1633 had repercussions for the science of plants at Oxford. In the Vatican, the Inquisition found an old man guilty of heresy, whose crime was to subscribe to Copernican evidence that the earth moves around the sun, which was contrary to Catholic dogma. Galileo Galilei's punishment

was house arrest for the rest of his life.[48] In Palermo, Paulo Boccone was born into a wealthy family. He was to become court botanist to the Medici dynasty, a Cistercian monk and a highly regarded natural historian. In London, Samuel Pepys was born to a Fleet Street tailor's wife. Pepys eventually entered Cambridge University, and became well known for documenting Restoration society.

Galileo, an experimentalist, inspired the founders of the Royal Society, which in turn influenced how the Garden was stewarded in its early decades. Pepys, always a curious observer, visited the Garden and, as the society's president, helped ensure that it was part of the scientific establishment in late seventeenth-century England. Boccone, a recorder of nature, was associated with the introduction from Sicily of the eponymous Oxford ragwort, *Senecio squalidus*, whose biology has fascinated generations of botanists.

Collections for the curious

For hundreds of years before the foundation of the Botanic Garden, menageries or gardens across Europe, which boasted eye-catching, rare and unusual organisms, reflected the wealth, power and prestige of their owners. Living collections might be augmented by cabinets of curiosities, preserved collections of the singular, the dramatic and the elusive. One of the largest and best-known of these in England, which combined living and dead organisms with artefacts, was constructed by the John Tradescants in the early seventeenth century.

John Tradescant the elder was a traveller and gardener to the nobility, such as the secretary of state Robert Cecil, the high-profile courtier George Villiers and latterly King Charles I. As a traveller, he explored Arctic Russia, North Africa, the eastern Mediterranean and France. John Tradescant the younger succeeded his father as head gardener to Charles I, making journeys to the Americas, from where he introduced many novel plants to English gardens, for example the tulip tree (*Liriodendron tulipifera*) and the bald cypress (*Taxodium distichum*).[49] At Lambeth in London, the Tradescants constructed a public cabinet of curiosities, known as the Ark, and a garden which attracted visitors from across Europe. The collection's 1656 catalogue presents an eclectic mix of objects, assembled with at least one eye on the fickle, fee-paying public, rather than a coherent natural history collection.[50] As might be expected, the Tradescants' living collection included the mandrake.[51]

above **Portrait of John Tradescant the elder**, from John Tradescant the younger's *Musaeum Tradescantianum, or, A Collection of Rarities Preserved at South-Lambeth neer London* (1656). Bodleian Library, Sherardian Library of Plant Taxonomy, Sherard 163(2).

opposite **Illustration from *The Tradescants' Orchard***, a seventeenth-century bound manuscript of fruit paintings, found in Elias Ashmole's collection. Oxford, Bodleian Library, MS. Ashmole 1461, fol. 37r.

The pilot plum ripe
July the 24

When Tradescant the younger died in 1662, Elias Ashmole added the Ark, but not the Tradescants' garden, to his own substantial collection.[52]

Approximately forty years after Bobart the elder was appointed, the university acquired Ashmole's collection, one of the premier cabinets of curiosities in Europe. It became the foundation of the Ashmolean Museum in Oxford.[53] With Ashmole's gift, the university had the potential to create something akin to Bacon's utopian foundation, Salomon's House, with a 'universal museum' at the centre of an education and research institute.[54] The Ashmolean's first keeper, Robert Plot, held to this ideal, but with his departure from the post in 1690 the vision faded, and these functions evolved separately rather than through mutual collaboration across the university.[55] The university did not encourage Ashmole's collection and the Botanic Garden to work together, although both Plot and his successor, the naturalist Edward Lhwyd, held Jacob Bobart the younger in high regard for his knowledge of botany and horticulture.

The Royal Society changed the notion of a collection from something to inspire awe to a place where science could be conducted and the world understood. To further this aim, it created a 'repository' of experimental equipment and natural rarities that would have 'considerable Philosophical and Usefull purposes'.[56] Within decades, the ideal of the repository as a collection of evidence was subverted as mere curiosities multiplied, for example, 'a piece of a BRANCH naturally shaped like a *Penis* with a pair of *Testicles* annexed'.[57] With the collector *extraordinaire* Hans Sloane playing a leading role in the society's affairs from 1693, little could be done to reform the repository. However, on Sloane's death, the society quickly gave the repository's contents to the British Museum, which coalesced around Sloane's personal effects.[58]

Like the university, the Royal Society realized that a universal collection that provoked questions about the world, rather than merely preserved bits of it, was expensive and ultimately unachievable. Moreover, the seventeenth-century scientific revolution was a fragile creature in the university, and of only marginal concern compared to the great theological debates of the day, as it was to prove at other times during the next four centuries.[59]

The Garden in the seventeenth century

At the most mundane level, the Botanic Garden was a place where medicinal plants could be displayed with their correct names. For

Seventeenth-century oil portrait by an unknown artist, traditionally believed to be of Jacob Bobart the elder. University of Oxford, Department of Plant Sciences.

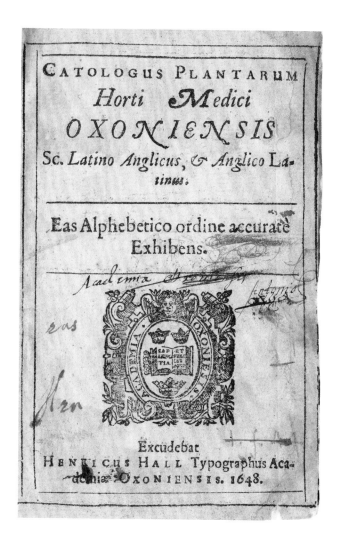

example, in 1658 the physician who 'be puffed up with vain perswasion of his own abilities, and shall think because he hath the title of Doctor he may be as idle as he please, and slight the study of Simples [fundamental ingredients of medicines]'[60] was warned that the Botanic Garden could teach him something. By the late seventeenth century, Thomas Baskerville argued the Botanic Garden was 'of great use & ornament, prouving serviceable not only to all Physitians, Apothecaryes, and those who are more imediately concerned in the practise of Physick, but to persons of all qualities seruing to help y[e] diseased and for y[e] delight & pleasure of those of perfect health'.[61] Yet, the physician Thomas Sydenham, who had studied medicine at Oxford, was less sanguine. He had a poor opinion of universities generally, and Oxford in particular, as places to learn practical medicine: 'one had as good send a man to Oxford to learn shoemaking as practicing physick'.[62]

A botanic garden as a medical teaching resource may have been what Danvers and the university had in mind. However, seventeenth-century lists of plants growing in the Garden paint a picture of their being grown for purposes as diverse as introducing novel plants to Britain from Europe and the Americas; testing the best conditions for growing plants; and supporting the research of Robert Morison, the Regius Professor of Botany, who had been appointed in 1669.[63]

Moreover, the Bobarts had to make money to keep the Garden running as the university funded only their salaries. They resorted to imaginative approaches to funding, including selling produce from the Garden at Oxford markets and supplying unusual plants and seeds to wealthy patrons such as Mary Somerset, Duchess of Beaufort, in the late seventeenth and early eighteenth century. The university provided resources for the Garden's maintenance only when the affairs of the English botanist William Sherard, founder of the Sherardian chair in the university, were settled in 1734.

Despite its academic role, Bobart the elder laid the Garden out according to the highly formalized fashions of the day. It was to look

Title page of the anonymous *Catologus plantarum horti medici Oxoniensis* (1648), the first catalogue of the Oxford Botanic Garden, which is traditionally attributed to Jacob Bobart the elder. Bodleian Library, Sherardian Library of Plant Taxonomy, Sherard 23.

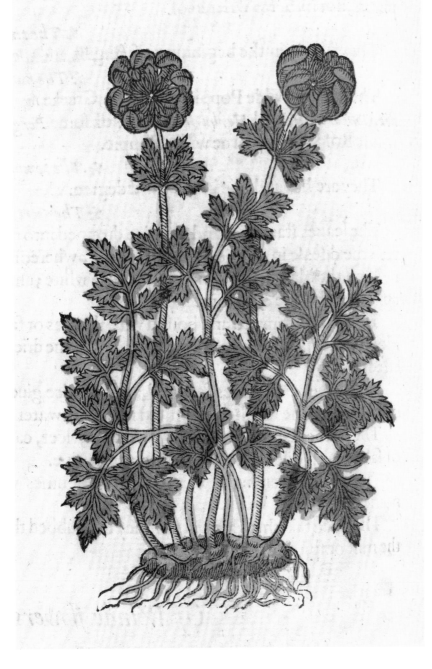

2 *Anemone coccinea multiplex.*
Double scarlet Winde flower.

beautiful. It was a place to be shown off and to reinforce the power and reputation of the university. Distinguished visitors such as the Prince of Orange (the future King William III) were paraded through it, and gentlemen such as John Evelyn and Elias Ashmole had nothing but praise for it.[64] There were less supportive voices. In 1664 the French physician Samuel de Sorbière dismissed the Garden as 'small, ill kept, and more like an Orchard than a Garden'.[65] Five years later, Cosimo III de' Medici, Grand Duke of Tuscany, thought the Garden 'scarcely deserves to be seen' from the 'smallness of its site, irregularity, and bad cultivation'.[66]

The Garden's site, and the limitations it places on botanical teaching and research, has been a recurrent theme through the history of botany at Oxford. At the end of the eighteenth century, the professor of botany John Sibthorp complained to his students that the Garden was 'greatly inferior in Magnificence & Splendour to those supported by Royal Expenditure [e.g., Kew, Jardin du Roi]'.[67] A century later, another professor of botany, Sydney Vines, also complained about the state of the site.[68] In the 1950s the study of botany moved away from the Garden to purpose-built facilities elsewhere in the city. By the start of the twenty-first century, even this site was insufficient for the botanical ambitions of the university.

Nevertheless, the Botanic Garden was the focus of scientific investigations of plants in the university from the 1640s until 1951, when the Department of Botany moved out of the Garden's confines to South Parks Road. Plants within the Garden were used for teaching seventeenth-century physicians, but the first superintendents implicitly rejected the notion that botany was merely the handmaiden of physic; they held that plants had an interest that was independent of their food and medicinal uses. Moreover, within the meagre resources of the Garden, the botanists and gardeners who worked there aspired to acquire new knowledge about plants. The methods of the nascent Royal Society, which demanded objective evidence – *Nullius in verba* (Take nobody's word for it) – had taken root.

Copper engraving of a plan of Oxford Botanic Garden from David Loggan's *Oxonia illustrata* (1675). Oxford, Bodleian Library, Vet. A3 b.21.

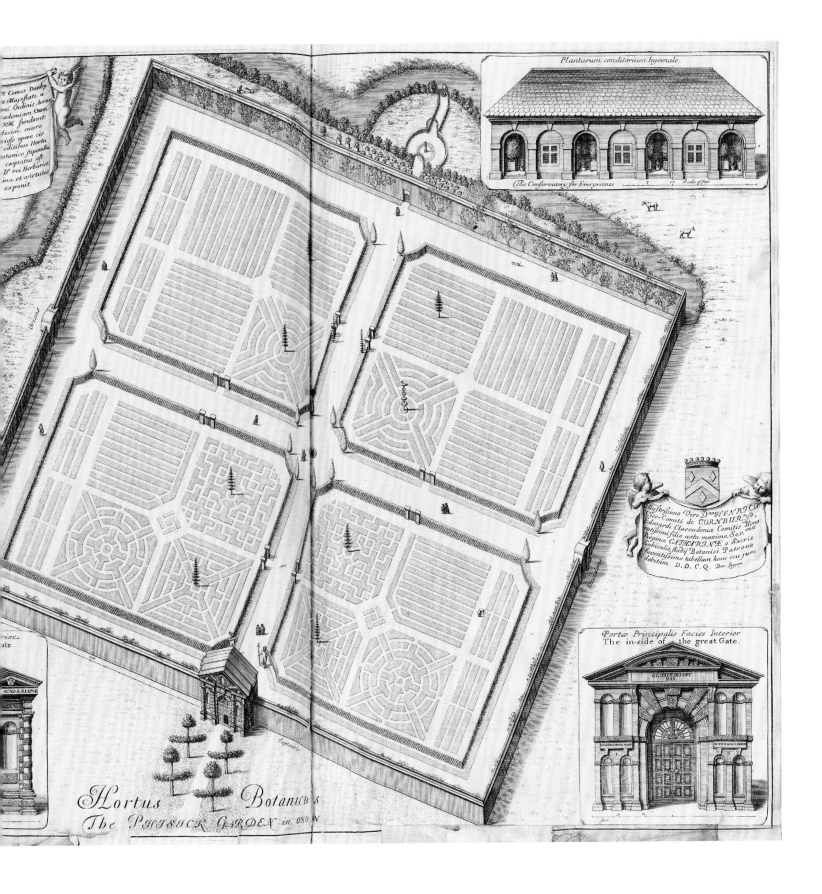

Plantarum conditorium hyemale.

The Conservatory for Evergreens

Scale of feet

S Comes Danby
r Majestati
mi Ordinis Auras
ademiam Oxon
ISM fundavit
tecim muro
vioso opere cir
editibus Hortu
otanico stipenbia
cooptatus es
D. rei Herbaria
ina et virtutes
exponit

Illustrissimo Viro Dno HENRICO
Vice-Comiti de CORNBURY,
Eduardi Clarendoniæ Comitis Hono
ratissimi filio natu maximo, Ser. mæ
Reginæ CATHARINÆ a Sacris
Cubiculis, studij Botanici Patrono
faventissimo tabellam hanc ceu jure
debitam D.D.C.Q. Dav. Loggan

Portæ Principalis Facies Interior
The in-side of the great Gate.

GLORIÆ DEI OPT
MAX.

Hortus Botanicus
The PHISICK GARDEN in OXON

Henry Danvers (1573–1644) [69] was a Wiltshire-born soldier and administrator who was amply rewarded for his efforts on European battlefields during a particularly volatile period of the late sixteenth and early seventeenth centuries. He fought in the Netherlands, France, Spain and Ireland in the service of men such as Maurice, Prince of Orange, Robert Devereux, 2nd Earl of Essex and Charles Blount, 8th Baron Mountjoy.

In 1594 Henry and his elder brother Charles murdered a member of a rival Wiltshire family, and were helped to escape to France by Henry Wriothesley, 3rd Earl of Southampton. The brothers entered the service of Henri IV of France and were pardoned by Elizabeth I in 1598. Charles took part in the rebellion of the Earl of Essex, and was beheaded three years later, but Henry enjoyed the favour of both James I and Charles I. James made him Baron Danvers in 1603 and life governor of Guernsey in 1621, and Charles made him Earl of Danby in 1626. Henry's younger brother John was one of the regicides who signed Charles I's execution warrant.

Danvers was immensely wealthy, but his retirement at Cornbury Park, Oxfordshire, was marked by infirmity. Unmarried and childless, he ensured that he would be remembered by a benefaction that established the Oxford Botanic Garden.

Jacob Bobart the elder (*c.*1599–1680),[70] the 'Germane Prince of Plants', was born in Braunschweig, modern Germany, and became a soldier before he settled in Oxford. In 1642 Bobart was made the first superintendent of the Botanic Garden. Little is known about him other than he was a tall, strong, literate, eccentric man of integrity with a penchant for topiary.[71] Contemporary portraits show a long-bearded, rather severe-looking man who might 'hold his own among the dons of the university',[72] but who was the butt of town-and-gown wits. Bobart is credited with the authorship of the *Catalogus plantarum horti medici Oxoniensis* (1648), the first catalogue of plants in the Botanic Garden.

Bobart married twice, and had at least ten children. His eldest son, Jacob, became his successor at the Garden.[73] When he died, Bobart was financially secure, owning leases for the profitable 'Greyhound Inn and meadow' and houses at Smythgate (north Catte Street). He also owned a house on George Lane (George Street) and bequeathed more than £115 (*c.*£13,000 in 2020) to his daughters. The Swedish botanist Carl Linnaeus commemorated Bobart and his son Jacob with the generic name *Bobartia* for a group of South African plants in the iris family.

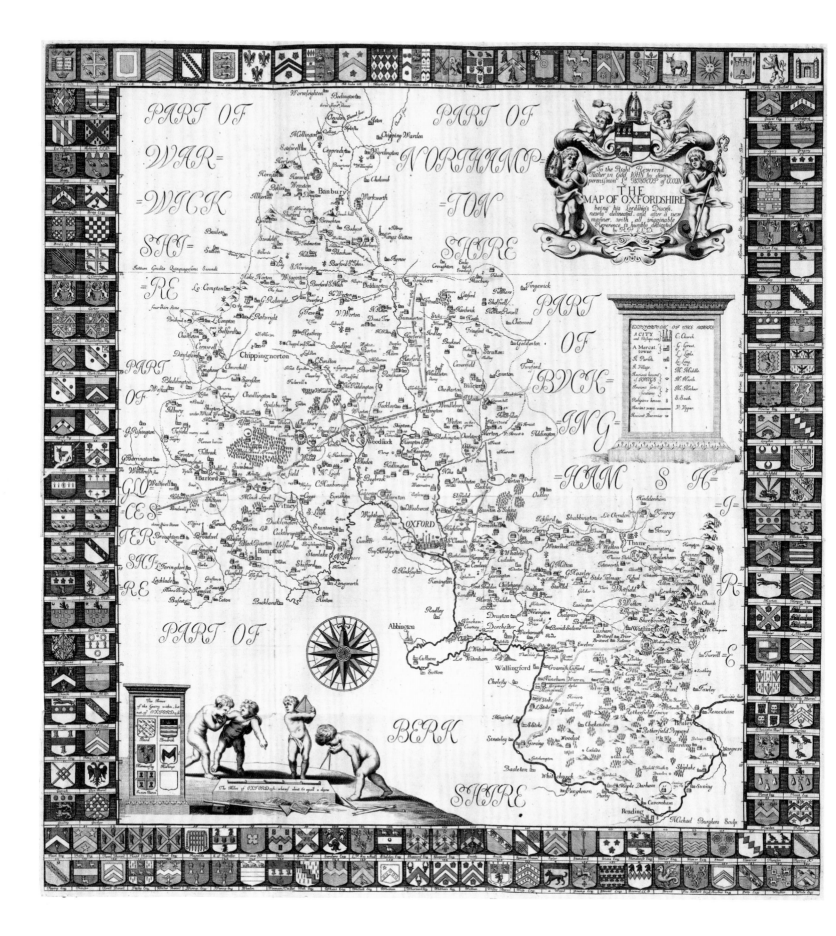

PART OF WAR=WICK SHI=RE

PART OF NORTHAMP=TON SHIRE

PART OF BVCK=ING=HAM SH=IRE

PART OF GLO=CES=TER SHI=RE

PART OF BERK SHIRE

THE MAP OF OXFORDSHIRE

To the Right Reverend Father in God JOHN by devine permission LORD BISHOP of OXON THE MAP OF OXFORDSHIRE being his Lordship's Diocess, newly delineated, and after a new manner, with all imaginable Reverence is humbly dedicated by R.D.

Banbury

Deddington

Chipping norton

Woodstock

Burford

Witney

Bampton

OXFORD

Thame

Watlington

Wallingford

Henley

Abbington

Reading

Michael Burghers Sculp.

Robert Plot, FRS (1640–1696),[74] was professor of chemistry at Oxford university and the first keeper of the Ashmolean Museum, England's premier seventeenth-century cabinet of curiosities. Born in Kent, Plot entered Magdalen Hall (now Hertford College), Oxford in 1658, and over the next three decades forged and burnished his reputation as the 'learned Dr Plot'. He was fascinated by experimental approaches to understanding the natural world, which were emerging at Oxford and coalescing around natural philosophers such as Robert Boyle. In 1677 Plot published *The Natural History of Oxford-shire, being an Essay toward the Natural History of England*. Plot's aim was clear: he wanted to provide rational, detailed descriptions of objects and phenomena, and he believed in the value of experimentation to understand the natural world.

Plot brought to the Ashmolean his enthusiasm and high standards of curation, but also an acquisitiveness that was 'to disgust some of his fellow antiquarians'.[75] By 1689 Elias Ashmole appears to have lost confidence in his curator: 'he does no good in his Station, but totally neglects it wandering abroad where he pleases'.[76] Plot was forced to give up the post in 1690 when he decided to marry.

Map of Oxfordshire, from Robert Plot's *The Natural History of Oxford-shire, being an Essay toward the Natural History of England* (1677). Oxford, Bodleian Library, Douce P 19, insert map.

1
a *b*

2

3

II

2 STEM

Collections

All plant biologists rely on data derived from collections of one form or another. These include collections of dried plants from herbaria; living plants from gardens or experimental trials; seeds or cell cultures from germplasm banks; DNA, RNA and protein sequences from databases; experimental data sets; and books, journals and manuscripts. Collections allow comparison between the samples they contain, and for these samples to be compared with those used in the past. They give current researchers confidence in data collected by their predecessors, while providing data that may prove useful to their successors. In short, collections are stores of evidence – empirical data (the facts) – from which information can be distilled, and scientific ideas generated and tested. They represent intellectual capital which evolving philosophical, technological and analytical frameworks can use to investigate novel questions about plants.

Some of the University of Oxford's botanical collections are old. They provide data about extant plants that extends back at least four centuries.[1] They are also rich in the quantity, quality and geographical extent of the data they contain. Complementing the transient, living exemplar plants in the Botanic Garden is the Herbaria's permanent preservation of these plants. At a certain point the university started to maintain experimental collections, where plants grown under controlled conditions could be rigorously compared with each other. Current plant science research in the university would be impossible without the exchange of data between national and international collections of seed banks, mutants, cell lines and, of course, sequence and trait (plant characteristic) databases.

The research undertaken by plant scientists at Oxford today contributes hundreds of thousands of data points to global databases that can be used by other researchers. Such innovations are barely twenty-five years old. In its herbaria, gardens, libraries and archives, Oxford has a much longer history of stewarding data for future generations.

Hand-coloured metal engraving of a female cannabis plant (*Cannabis sativa*) from Elizabeth Blackwell's *Herbarium Blackwellianum* (1760). Bodleian Library, Sherardian Library of Plant Taxonomy, uncatalogued, v.4, t.322a.

HEMP

Hemp[2] (*Cannabis sativa*) is a tall, wind-pollinated annual herb with separate male and female plants. Tough fibres isolated from the plant's stem constituted the ships' rigging that helped some Western nations control vast areas of the globe, while its oil-rich seeds fed and fuelled parts of Asia. The psychoactive drug tetrahydrocannabinol, concentrated in minute glands on the flower buds, is used in modern medicine and worldwide for recreation.

Over thousands of years, humans have consciously, and unconsciously, modified hemp through cultivation. Field-bound hemp has also exchanged genes with wild hemp, creating complex patterns of variation in the plant's appearance. Whether the drug, food, fibre and oil all come from the same species in the genus *Cannabis* or from different species has been a matter of argument for centuries.

If we accept that humans use only one species, *Cannabis sativa*, two subspecies can be recognized, each of which has its wild and cultivated forms. The stems and fruits of one subspecies, ssp. *sativa*, have been used for fibre, food and oil production, while the flowers and leaves of the other, ssp. *indica*, have been used for drug production. The crux of the scientific discussion is accounting for, and understanding, the complex patterns of variation in a plant that has been intimately associated with humans for thousands of years. However, hemp science is more than just giving the plant a name; it is also about understanding the plant's biology, including its ecology, genetics, physiology, biochemistry and evolution.

Hemp plants growing in the Botanic Garden in the 1660s, and added to the herbarium of Bobart the elder, were called either 'Cannabis mas' (male cannabis) or 'Cannabis faemina' (female cannabis).[3] At a time when sexual reproduction in plants was unknown, the names were applied to the wrong genders. The robust, biologically female plants were called 'male cannabis', while the delicate, biologically male plants were called 'female cannabis'. Similar confusions between morphology and gender can be seen in other plants with separate male and female sexes in this period.[4]

Answers to fundamental questions in plant biology about gender and improving our understanding of hemp biology come from centuries of formal and informal research.[5] An essential part of this process has been the synthesis of data from global collections of dried specimens, experimental trials, and DNA and other sequence databases, as well as from libraries, archives and collections of unwritten traditional knowledge.

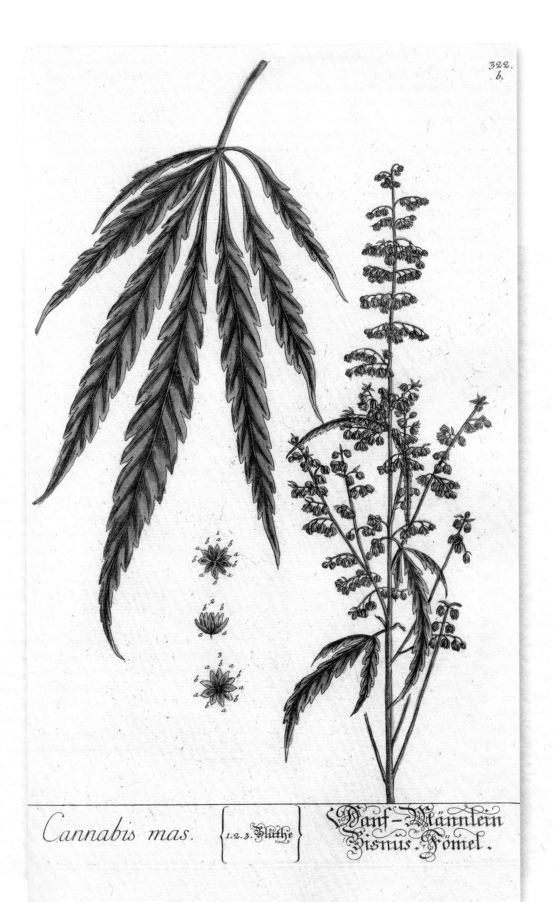

Cannabis mas. [1.2.3. Blüthe] Hanf-Männlein Hisnus. Fömel.

Hand-coloured metal engraving of a male cannabis plant (*Cannabis sativa*) from Elizabeth Blackwell's *Herbarium Blackwellianum* (1760). Bodleian Library, Sherardian Library of Plant Taxonomy, uncatalogued, v.4, t.322b.

The changing value of collections

When the Tradescants' Ark of preserved objects arrived at Oxford University in the late seventeenth century, it had no role in botany – it was a curiosity from another era – while the Tradescants' living collection at South Lambeth was a shadow of its former self.[6] By 1710, Joseph Addison, a fellow of Magdalen College and co-founder of *The Spectator*, was parodying such collections as the efforts of 'a sort of learned men, who are wholly employed in gathering together the refuse of nature'.[7] Yet, the first two keepers of the Ark in Oxford, Robert Plot and Edward Lhywd, were passionate about using the collections to further global knowledge of the natural world, including botany.

The slow evolution of the collections from a cabinet of curiosities to a carefully curated and labelled group of objects that could be used to address scientific botanical questions began at Oxford in the latter half of the seventeenth century. The change came with the Bobarts' expansion of the living collection at the Botanic Garden, as they privately engaged in creating a library of dried preserved plants – a herbarium – from specimens in the Garden and from the countryside surrounding the city.[8]

The mid-sixteenth-century professor of botany at the University of Bologna, Luca Ghini, probably originated in Europe the technique of preserving flattened plants by drying.[9] One of the earliest references to a herbarium is that made by one of Ghini's pupils: 'I never sawe it [*Lysimachia maritima*] in England savinge onelye in Master Falkonner's boke and that had he brought out of Italy except my memory do fayle me.'[10] However, it was only when the cost of paper fell in the late seventeenth century that herbaria became common scientific tools and desirable objects for the 'curious'.[11] Among the curious was Samuel Pepys who, on 5 November 1665, was surprised by John Evelyn's *hortus hyemalis* ('winter garden') with 'leaves laid up in a book of several plants kept dry, which preserve colour, however, and look very finely, better than any Herball'.[12]

The Bobarts' reason for creating a leather-bound, elephant-folio book herbarium is unknown. Perhaps they realized that the 2,800 specimens it contains provided important links with catalogues of plants growing in the seventeenth-century Garden.[13] Certainly, by the early eighteenth century, the Bobarts' novel collection was well known, at least in Oxford:

Herbarium specimen of cocoa (*Theobroma cacao*) collected in Jamaica, from William Sherard's personal herbarium, pressed on the drying sheet (as shown in the cocoa specimen on p. 41). Oxford University Herbaria, Sher-1580.

Thy Hortus Siccus still receives:
In Tomes twice Ten, that Work immense!
By Thee compil'd at vast Expence;
With utmost Diligence amass'd,
And shall as many Ages last.[14]

In the mid-nineteenth century, the Oxford chemist and botanist Charles Daubeny, fifth Sherardian professor, was busy rectifying decades of neglect of the Garden and its living collections.[15] When in 1853 he formally accepted, on behalf of the university, the vast herbarium of the British collector of botanical specimens, Henry Fielding, he believed that collections of living and of dried plants were complementary teaching and research tools. He argued that living collections encouraged people's interest in plants, and provided access to a large amount of seasonal material, from a limited array of species, for teaching, while a limited

The impression of a specimen of cocoa (*Theobroma cacao*) on a sheet of drying paper used as the cover of one part of Sherard's *Pinax*. Bodleian Library, Sherardian Library of Plant Taxonomy, MS. Sherard 56, endpaper.

amount of material from a great diversity of plants was available all year round in a 'dried garden'.

When Daubeny negotiated the acquisition of the Fielding collection, he secured a £900 endowment (c.£72,000 in 2020) from Fielding's wife, Mary, to pay a curator, but only patchy curatorial care was secured. The first Fielding Curator, twenty-year-old Maxwell Masters, was appointed in 1853; he resigned after about three years to teach botany at St George's Hospital Medical School in London before eventually taking on the editorship of *The Gardeners' Chronicle*. In 1886 a twenty-six-year-old German botanist, Selmar Schönland, arrived for a three-year stint as Fielding Curator before emigrating to South Africa, where he founded the botany department at Rhodes University.[16] The post was then vacant until the Oxford-based pharmacist George Claridge Druce became honorary curator in 1895; he remained in post until his death thirty-seven years later.[17]

Oversight of the Oxford botanical collections was marked by two notable events. Until the 1880s, the books, archives and manuscripts associated with the living and preserved botanical collections were accommodated by the professor of botany. Consequently, when the third Sherardian professor, John Sibthorp, died in 1796, much of the archive was in his home in Cowley House (now part of St Hilda's College). His relatives, wanting the space, sold much of the archive as waste paper to Oxford tradesmen.[18] Similarly, in 1942, when the longest-serving superintendent of the Garden, William Baker, was required to retire after fifty-four years' service, he destroyed most of the Garden's planting records in an apparent fit of pique.[19] At best, such events reflect a lack of vision and confidence in the long-term value and scientific role of Oxford's botanical collections.

During the twentieth century, disagreements about the important or interesting questions in botany (and biology more generally), together with the high cost of research and diminishing resources to conduct such research, often set the stewards of Oxford's botanical collections at odds with professors, university authorities, collection users and research funders. This is not unique to Oxford – it happens at a national, even international, level, with botanical collections amalgamated, stored or discarded as methods of scientific investigation evolve.

Botanical collections as part of the university's intellectual heritage, capital and environment must adapt to the landscape of modern plant

previous pages **Named cultivars of auricula (*Primula auricula*)**, from the book herbarium of Jacob Bobart the elder, the foundation of the university's botanical collections. The specimens were probably collected in the Botanic Garden in the 1660s. Oxford University Herbaria, BSn-A33r.

sciences. Today, responsibility for the university's botanical collections is shared by three institutions within it: Oxford Botanic Garden and Arboretum; the Bodleian Libraries; and, in the case of the Oxford University Herbaria, the Department of Plant Sciences.

Living collections

Living collections have important roles in botanical teaching and research. During the past three decades, such collections have reinvented themselves to inform us about the importance of plants in our daily lives and the threats they face in their natural environments. The Garden has, to a greater or lesser extent, played these different roles in the exchange of botanical information since it started to be planted in the 1640s.[20]

The Bobarts were part of a community of city and college gardeners exchanging techniques, observations and plants, and taking advantage of mystical, practical and experimental interests in botany. Then, as now, visitors expected to find exotic plants in the Garden, so the Bobarts developed conservatories and stoves to cultivate their more tender specimens. The simple catalogue of approximately 1,400 plants growing in the Botanic Garden, published anonymously in 1648, would be of little scientific use today had it not been for the physical specimens the Bobarts collected, which survive in the university's Herbaria.[21]

By matching names in the publication with the handwritten names the Bobarts attached to the specimens, one can be sure of what plants were actually grown. In addition to a small collection of medicinal plants, there were novelties such as introductions from the Americas, for example, the 'humble plant' (*Mimosa*), whose leaves folded when touched, and variegated plants collected from the wild and from college gardens. Horticultural favourites included morphological and colour forms of anemones (*Anemone*), oriental hyacinths (*Hyacinthus orientalis*) and wallflowers (*Erysimum cheiri*). By the end of the seventeenth century, the walls of the Bobarts' Garden were covered with trained trees and shrubs such as apples (*Malus*), cherries (*Prunus* subg. *Cerasus*), medlars (*Mespilus germanica*), peaches (*Prunus persica*), pears (*Pyrus communis*), plums (*Prunus* subg. *Prunus*) and quinces (*Cydonia oblonga*).[22] The seeds and plants that were stocked in the Bobarts' Garden came from their own collections made in the wild, gifts from friends and colleagues, and exchanges between fellow gardeners. Some plants may

Herbarium specimens of the myrtle family (Myrtaceae), including types in red folders, arranged in Oxford University Herbaria. Oxford University Herbaria.

even have been purchased, as gardens were judged on the number of unusual specimens they contained.

We know that the Garden was formally laid out and quartered by yew hedges, but, despite all the information we have about what it contained, we know very little of how the plants were arranged. They might have been organized according to medicinal or other utilitarian uses, or geographically according to the plants' origins. During the eighteenth century, the taste for formal gardens declined, and the third Sherardian professor, John Sibthorp, redesigned the beds and planting on a broadly geographical basis, with Britain and Europe east of the north–south path, and North America and Asia to the west. Sibthorp's Garden does not appear to have been manicured: the plants showed 'their Natural Growth, neither disguised nor distorted by Art', and their detailed arrangement reflecting his teaching interests.[23]

In the nineteenth century, Charles Daubeny and the Garden superintendent William Baxter radically redesigned the Garden with picturesque and curved beds. In the 1880s the seventh Sherardian professor, Isaac Bayley Balfour, swept away Baxter's beds and planted 'order beds' according to the then current taxonomic system of Kew's George Bentham and Joseph Hooker. During the twentieth century, academics argued the merits of different classification systems, but in the early 2000s the director of the Garden, Timothy Walker, rearranged the beds according to a plant classification based on DNA data. Elsewhere in the Garden, a rock garden was added in the 1920s, and a herbaceous border laid out in the 1940s. Horticultural innovation in the Garden has continued in recent years, including radical designs such as the Merton Borders, whose aim was to combine visual impact, low-input establishment and cost-effective management.

In the 1730s the Garden entered the race for horticultural one-upmanship; two modest architectural conservatories were built for collections of exotic rarities, such as coffee (*Coffea arabica*), tea (*Camellia sinensis*), cotton (*Gossypium*), sugar (*Saccharum officinarum*) and pineapple (*Ananas comosus*), and tender ones, especially those from the Royal Apothecary James Sherard's garden at Eltham in Kent.[24] Growing exotic plants came at a price: between 1735 and 1754, approximately 40 per cent of the Garden's annual recurrent budget was spent on keeping them alive through the costs of heating and maintaining conservatories. By the end of the century the conservatories proved to be more decorative than practical.

Daubeny's desire to grow an exotic water lily led to the greatest transformation of glasshouses in the Garden. In 1851 the Lily House tank, heated by hot water flowing through iron pipes, became the centrepiece of Daubeny's glasshouse complex. The Amazonian water lily (*Victoria amazonica*) flowered two years later. However, the foundations of the tank were poor, and continual maintenance became such a financial burden on the Garden that Daubeny eventually funded it out of his own pocket.[25] By the early twentieth century the problems had been resolved, but growing tender exotic plants has meant that the glasshouses have had to be replaced approximately every forty years since the mid-nineteenth century.

A scientific collection requires more than just the housing of an object, whether it is living or dead. For specimens to have scientific value, they must be associated with data about when and where they were collected, and multiple examples from across a species' distribution must be represented. Premiums are therefore attached to unique specimens of known wild origin, rather than those that come from commercial sources or through exchange with other collections.

The Garden in central Oxford has little prospect of creating a collection based on many representatives of a single species: the space is too small and the site is too valuable to be used for such a purpose. Visitors usually want to see many different things, while researchers are often interested in differences between many different examples of rather similar things. The tension is particularly stark with collections of trees.

Trees have always been part of the Garden landscape, but in the 1840s Daubeny actively began to build a collection of cone-bearing trees (gymnosperms) for economic and aesthetic reasons.[26] The pinetum, planted between the Garden's western wall and Rose Lane, included trees such as the monkey puzzle (*Araucaria araucana*), the Nootka cypress (*Cupressus nootkatensis*) and the Bhutan pine (*Pinus wallichiana*). It was designed to 'select at once the hardiest and the most ornamental'[27] species, but an arboretum on the Garden site failed, although two of the original plantings remain.

The Garden had its arboricultural opportunity in 1963, when the university gave over part of the Nuneham Park estate, at Nuneham Courtenay, approximately eight kilometres south-east of Oxford. The nucleus of the Arboretum at Nuneham Courtenay is the pinetum started by Bishop Edward Venables-Vernon-Harcourt in 1835. Gradually, the area

Late nineteenth-century botanical specimens preserved in alcohol and used in teaching.
Oxford University Herbaria.

ARCEUTHOBIUM CRYPTOGAM
LORANTHACEAE

of the Arboretum has increased to approximately fifty-three hectares. Today, the Arboretum has a collection of mature conifers, especially from North America, within a historic landscape. The heights of the trees, particularly specimens such as the coastal redwood (*Sequoia sempervirens*), giant redwood (*Sequoiadendron giganteum*) and incense cedar (*Calocedrus decurrens*), which are among the oldest trees in the Arboretum, together with the kind of soil they need to grow, mean that it is impossible to grow them within the confines of central Oxford. Space at the Arboretum means that populations of a species can be planted rather than single specimen trees.

Herbaria

The collegiate university's collection of over one million herbarium specimens is concentrated in the Oxford University Herbaria within the Department of Plant Sciences. Other smaller, usually pre-nineteenth-century, herbaria are deposited in the Bodleian Libraries and in colleges such as Merton, Oriel and Wadham. In addition to the collections of the former Department of Botany, the Herbaria includes those from the former Department of Forestry and the vast personal collection of George Claridge Druce.[28] Its name is treated as a singular noun.

The Herbaria has its foundation in private collections made by the Bobarts before 1720 but, in 1728, the botanist and diplomat William Sherard bequeathed his personal herbarium to the university, thereby transforming its holdings into a collection of international significance. A law student at St John's College, Sherard had gone on to study botany in Paris and Leiden, establish an extensive network of personal and professional relationships across Europe, and fund plant collection expeditions in Europe and the Americas. In the late eighteenth century, James Edward Smith, founder of the Linnean Society of London, described Sherard's herbarium as 'the most ample, authentic, and valuable botanical record in the world'[29] after that of the Swedish 'father of taxonomy' Carl Linnaeus, which he had bought from Linnaeus' family.[30]

During the eighteenth century, the Bobart and Sherard herbaria were augmented with the collections of subsequent Sherardian professors, including Johann Dillenius and John Sibthorp, together with large donations such as the herbarium of Charles Dubois, cashier general of the East India Company. In 1850 Daubeny reported 43,812 specimens in the

Watercolour of a basket stinkhorn (*Clathrus ruber*), from Bruno Tozzi's *Sylva fungorum* (1724), part of a paper museum used by Johann Dillenius to study fungi. Bodleian Library, Sherardian Library of Plant Taxonomy, MS. Sherard 197, f.181r.

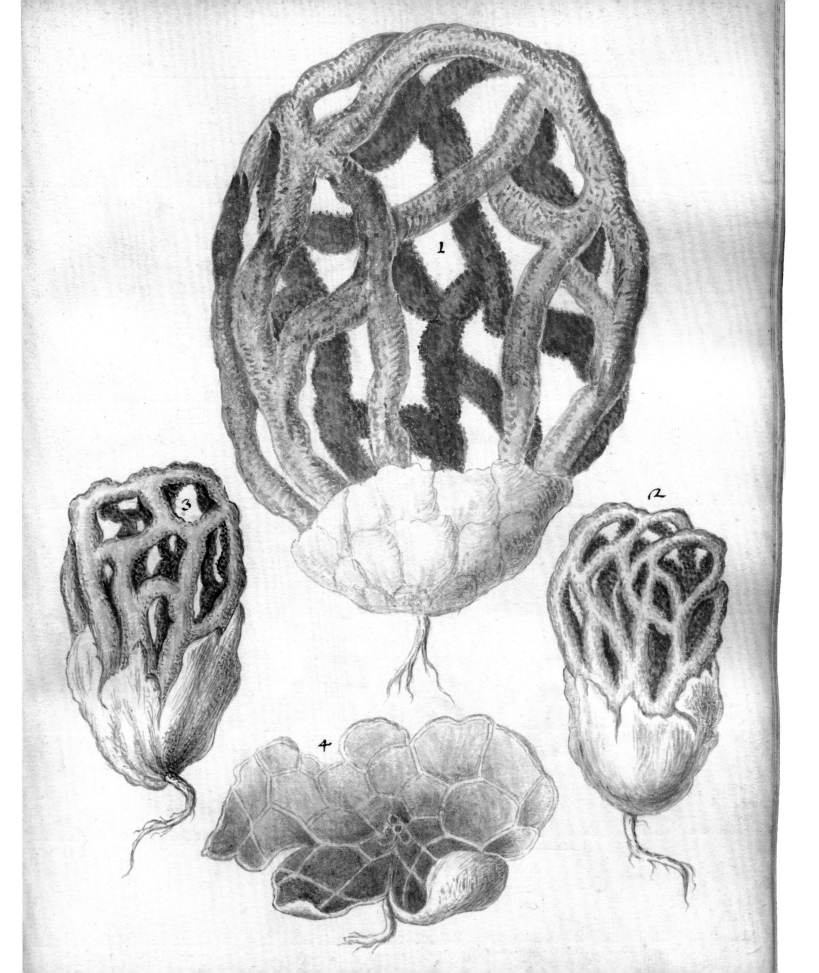

university's herbarium, but this appears to be an underestimate: there are approximately 80,000 pre-1750 collections in the Herbaria today.

In 1852 Daubeny convinced the university to accept the herbarium of Henry Fielding, which contained approximately 80,000 specimens. Fielding had spent much of his inherited fortune accumulating one of the largest personal herbaria in the Victorian world. Believing that the task of cataloguing the diversity of the world's plants was nearing its end, Daubeny praised the completeness of Fielding's collection with optimistic bravura:

> So large a portion indeed of its [the world's] surface has been ransacked to supply the contents of these cabinets, that it would seem to be a much shorter task for me to enumerate the deficiencies, than to recount the contents of the Collection.[31]

Over the next seventy years, at least 25,000 specimens were added to the herbaria through donation and the purchase of collections.[32] By the early twentieth century, the herbarium of the Department of Botany comprised about 200,000 specimens.

In the early twentieth century, in parallel with the development of the Department of Botany herbarium, herbaria were being created by the Imperial Forestry Institute, and privately by George Claridge Druce. The forestry herbarium, under the curatorship of the agronomist and tropical forester Joseph Burtt Davy, opened its doors in 1924.[33] Burtt Davy was concerned to ensure that the collection was suitable for teaching botany to colonial forest officers, and that it could help identify timbers that had (potential) economic value. During these early days, Oxford-trained forest officers were sending up to 8,000 specimens a year from across the British empire. A decade after it was founded there were some 54,000 specimens in the collection, the bulk of which were from anglophone Africa and Asia. After the Second World War the collection continued to grow rapidly, but its geographic focus broadened as researchers started to investigate plant diversity from across the tropics.

George Claridge Druce, honorary curator of the Fielding herbarium, began building his collection as a young man in the early 1870s. Until his death in 1932 he was consistently adding to it about 4,000 specimens a year. The collection, which the university reluctantly accepted on Druce's death, contained approximately 200,000 specimens, most of which

Selection of dried legume fruits associated with specimens in Oxford University Herbaria. Oxford University Herbaria.

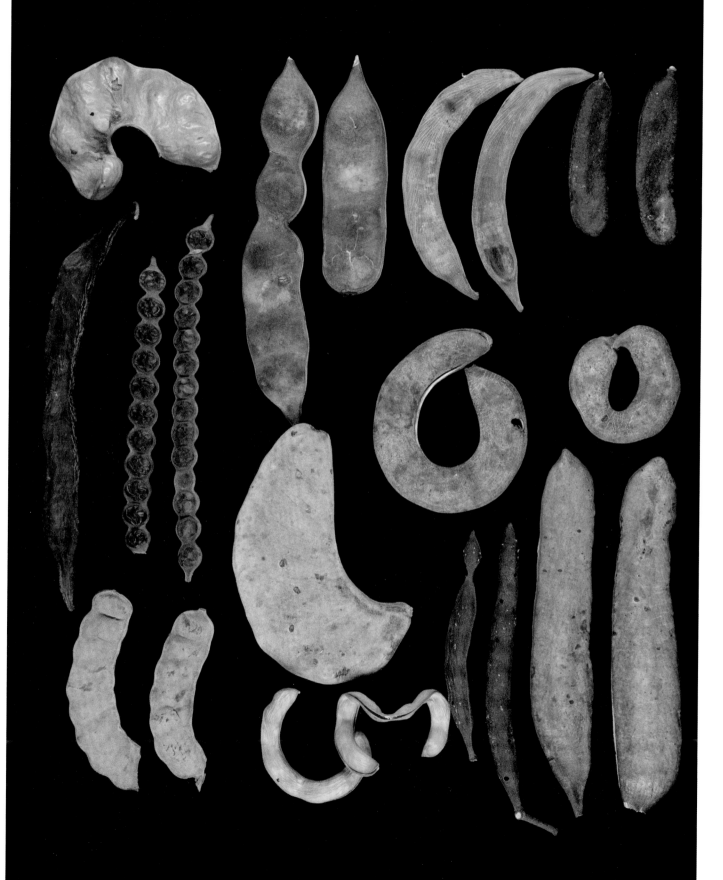

had been collected in the United Kingdom. It was the largest and most important collection of British plants in private hands.[34]

The 'House for Evergreenes', outside the Garden's north wall, which had been built in 1670, had been converted to accommodation for the professor and the university's collection of dried specimens in the early eighteenth century. In the 1780s it was demolished in a road-widening scheme, and the second Sherardian professor, Humphrey Sibthorp, had Cowley House built. John Sibthorp's death in 1796 saw the Eastern Conservatory in the Garden converted to accommodate the herbarium. When Charles Daubeny took over as professor, this space was needed for other purposes and the herbarium was evicted.

Daubeny created a 'Room for Seeds & Herbarium', which backed on to a garden shed, erected on the banks of the Cherwell. With the arrival of Fielding's specimens, the university's herbarium collections gained a new home on two floors of the converted Western Conservatory. The ladder between the floors was apparently 'so shaky' that the eminent Victorian botanist, the Reverend William Newbould, stated that, but for the 'strain it inflicted upon his nerves', Oxford would have been his place of residence.[35] By 1885, just before the Department of Botany was formally created, the herbarium was moved again – into the house Daubeny had built on the eastern side of the Danby Gate. Here the collection remained until it was finally moved to a new building on South Parks Road in 1951.

The South Parks Road site provided purpose-built accommodation for herbarium specimens. One room was to house the Department of Botany's and Druce's collections, the other the forestry collections. In their new home the forestry collections continued to grow rapidly as the specimens started to be used for research and teaching in ways that had not been seen since the early eighteenth century, and the work of the first Sherardian professor Johann Dillenius. However, each collection remained the bailiwick of a separate curator, with different policies and practices; for example, specimens in the different collections are mounted on different sizes of paper. In 1971 the herbaria of the Forestry and Botany departments merged under the curatorship of the tropical botanist Frank White. At last, the university had a single herbarium, albeit spread over two floors in the Department of Botany on South Parks Road. During the 1990s, the combined collections started to be called the Oxford University Herbaria.

Libraries

The library is the collection that is familiar to most plant scientists, although today it is usually through remote access to virtual documents rather than as a space containing physical objects. Until the seventeenth century, books were primarily the trophies and tools of the wealthy and of scholars. As private individuals became more interested in horticulture and the general populace became wealthier and more educated, they wanted plants to show off in their gardens, and information about plants and how to grow them. By the eighteenth century, a great wealth of books was being published, although it is not always clear at whom they were aimed.[36] Furthermore, publishing was aided by the fashion for having libraries in houses and the need to stock them.

The wealth of the university's botanical book and manuscript collections has been achieved partly through purchase, but most importantly through the generosity of benefactors who bequeathed their personal libraries and archives to Oxford. These individuals include the Bobarts and William Sherard; Sherardian professors such as Johann Dillenius, John Sibthorp, Charles Daubeny, Marmaduke Lawson and Sydney Vines; and herbarium curators such as Joseph Burtt Davy and George Claridge Druce. The library collections are the glue that hold the other botanical collections together. The interplay between the collections enables us to understand not only individual objects, but also how and why they came into the collection.

Consider the proof copy *Plantarum Dioscoridis Icones*, which John Sibthorp obtained in Vienna in the mid-1780s. The earliest surviving copy of Dioscorides' *De materia* is the magnificently illustrated *Codex Vindobonensis* (*c.*512 CE), which was made in Byzantium for the noblewoman Anicia Juliana and was part of the Imperial Library in Vienna by 1570. In the late eighteenth century, there was an ambitious project to reproduce all the illustrations in this manuscript. The project was never completed, but *Plantarum Dioscoridis Icones* is one of only two proof copies. Sibthorp used this as his field guide for his explorations

Spines of books in the personal library of William Sherard, a core part of the university's pre-1750 botanical collections. Bodleian Library, Sherardian Library of Plant Taxonomy.

of the eastern Mediterranean in the late 1780s,[37] and it became part of Sibthorp's personal library. The names it contains are important for interpreting the material Sibthorp brought back from his expedition, which was finally published as the *Flora Graeca* (1806–40), one of the world's rarest botanical books. *De materia*, the ultimate authority on plants for over 1,500 years, was still being used as a field guide in Greece in the 1930s.[38]

The forestry library created by the Oxford Forestry Institute, in its various guises, during the twentieth century was, by the end of the century, a peerless collection. Forestry books and journals were complemented by vast amounts of so-called grey literature – material not published in a conventional manner – which is rarely kept by libraries or the organizations that created it. This part of the library includes research reports, working papers, conference proceedings and reports produced by government departments, academics, business and industry. Such literature is traditionally treated with suspicion by academics but is now a rich resource for investigating long-term problems associated with environment change.[39]

Wood collection

A herbarium is a generalized plant collection containing representative parts of plants preserved for future research. In contrast, the Oxford xylarium contains only one sort of material – wood – and is therefore a highly specialized collection. As with the Herbaria, the xylarium is a node in an international network of wood collections.

The Oxford xylarium was established in 1924, with the foundation of the Imperial Forestry Institute, by the American wood technologist Carl Cheswell Forsaith, who was on secondment from Syracuse University in New York state. The first specimen added was a block of wood from the European lime (*Tilia* x *europaea*) donated by the university's School of Rural Economy, later to become the Department of Agriculture. The oldest specimen, a gift from Cambridge University, is an early nineteenth-century block of almond (*Prunus dulcis*) from the garden of Erasmus Darwin, the grandfather of Charles Darwin. Today the collection contains approximately 24,000 hand-sized samples, representing approximately one-fifth of the world's tree species from nearly 200 different countries. As might be expected, given the collection's origin as an imperial research tool, most of the samples

Nineteenth-century copy of an oil portrait of the first Sherardian Professor of Botany, Johann Dillenius, by an unknown artist. University of Oxford, Department of Plant Sciences.

come from former British territories. The xylarium is also biased towards economically important timbers.

With Forsaith's return to the United States, the collection became the responsibility of the wood anatomist Leonard Chalk. Under Chalk's guidance, the xylarium started to grow rapidly and to gain an international reputation for the quality of the research undertaken using it. Wood blocks were exchanged with other xylaria that began to be established worldwide. Specimens were donated by forestry officers and wood companies across the British empire, and samples collected through the fieldwork of Oxford-based researchers. Chalk's interest was the investigation of wood structure, and the use of anatomical characteristics for wood identification.[40] For such research he needed to understand how wood characters varied within and between species, which could be investigated only in a large, diverse collection of timber blocks. During the 1930s Chalk and his colleague in the Imperial Forestry Institute, Mary Margaret Chattaway, together with Samuel Record at the Yale University School of Forestry, and Bernard Rendle of the Forest Products Research Laboratory at Princes Risborough in the United Kingdom, laid the foundations for the investigation of wood anatomy.[41]

A testament to Chalk's efforts is that when he retired in 1963, nearly 60 per cent of the current collection had been acquired. Thin sections, mounted on microscope slides, made from over half of these samples, revealed the features essential to investigate wood anatomy. During the 1960s the focus of the xylarium shifted towards the mechanical properties of timbers and how these relate to wood anatomy.[42] By the mid-1970s, when the collection became the responsibility of the forest geneticist Jeffery Burley, interest in xylaria, which had been created primarily for anatomically based wood research, was in global decline.

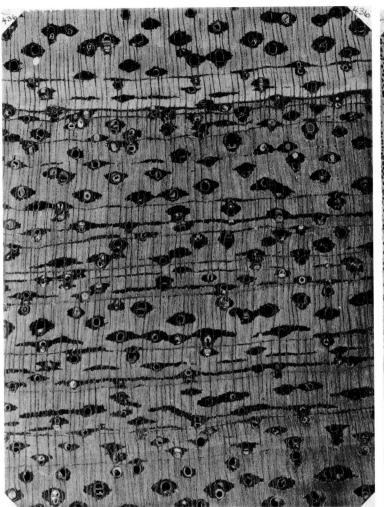
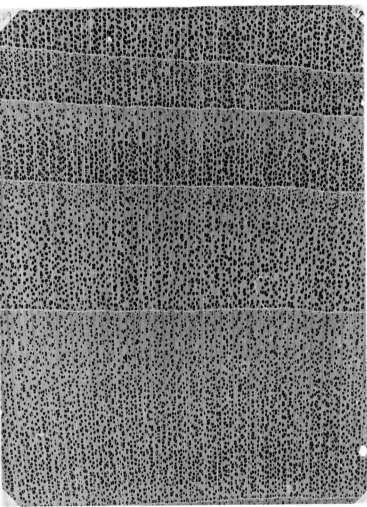

Experimental plots

Daubeny and Baxter realized that, to develop botany within the university, it needed to have the facilities to grow short-term collections of plants for experimental purposes. Within months of taking over the Garden in 1834, Daubeny had laid out an experimental garden outside its eastern wall.[43] Here he conducted agricultural experiments on barley, buckwheat and turnips. By 1850, it was proving too small, and Daubeny had new plans for the site – a water lily house (see above). Consequently, in 1852, he bought a piece of land off the Iffley Road in Oxford as an experimental plot, which he bequeathed to the university, who eventually sold it for development.

Images of transverse sections of wood blocks (magnified) used by Leonard Chalk for his research on wood structure in the 1930s:
left African pod mahogany (*Afzelia quanzensis*);
right American tulip tree (*Liriodendron tulipifera*).
Oxford University Herbaria, full plates 436 and 439.

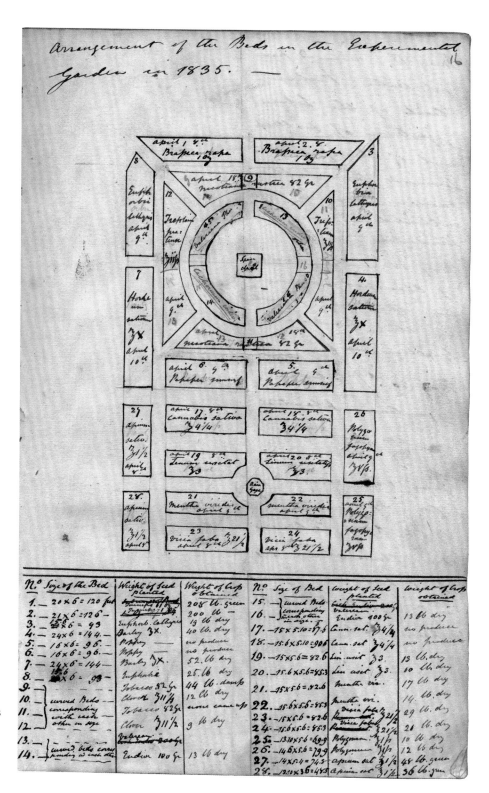

Charles Daubeny's 1835 sketch of the layout of the experimental garden he created in the Botanic Garden, which was lost when the glasshouses were built in the early 1850s. Bodleian Library, Sherardian Library of Plant Taxonomy, MS. Sherard 264, f.16r.

In the early twentieth century, roof-mounted glasshouses were added to the Department of Botany at the Garden. When the department moved to South Parks Road, the roofs were also used for experimental glasshouses, as was space at the Wytham Research Station, about five kilometres north-west of Oxford. These facilities provided control of experimental environments but did not enable large numbers of plants to be grown in natural environments. Oxford has never been endowed with either the number or the quality of experimental plots that were available at Cambridge University or at Rothamsted Experimental Station in Hertfordshire, which was founded in 1843 by the agricultural chemist John Bennet Lawes.

The importance of such facilities was recognized by Cyril Darlington in the 1960s, when he was Sherardian professor. Darlington established an experimental site, called the Genetic Garden, on the edge of the University Parks, near the Department of Botany, where the focus was on growing plants for genetic experiments, especially interspecific hybrids, variegated plants, domesticated crops and plants with variable chromosome numbers. In addition, Darlington's championing of the Arboretum was associated with the opportunities he could see for the site to be used for experimental plots, which the study of botany at Oxford so greatly lacked. Things changed with Darlington's departure: the Genetic Garden was abandoned, while the Arboretum had other priorities.

In the early 1970s, an alternative approach to the creation of experimental plots for ecological research was suggested in Oxford. Henry Colyear Dawkins, a lecturer in the Department of Forestry, systematically laid out permanent sample plots across Wytham Woods based on his experience of colonial forestry.[44] The idea was to use the natural woods as a living collection so that long-term changes in the composition and growth of the woods could be investigated. Over almost five decades, data from these plots have been recorded, which have provided detailed insights into change in forest landscapes.[45]

Herbarium specimen of the rare British plant herb-paris (*Paris quadrifolia*) collected by W.J.L. Sladen in Wytham Woods in 1953. Oxford University Herbaria, Elton_100.18.1.

William Sherard, FRS (1659–1728),[46] was one of the architects of early eighteenth-century European botany. The Swedish botanist and 'father of plant taxonomy' Carl Linnaeus described him as 'Botanicus Magnus'. To Linnaeus' student Fredrik Hasselquist, he was the 'Regent of the Botanic world', and to the editor of the letters of the Yorkshire naturalist Richard Richardson he was 'the Sir Joseph Banks of his day'.[47] Sherard's roles were as an identifier and nurturer of botanical talent, a maintainer of academic networks and the builder of one of the world's largest pre-Linnaean herbaria.

William Sherard was born in Bushby, the eldest son of a Leicestershire landowner. In 1677 he was awarded a fellowship at St John's College, Oxford, where he read law and graduated in 1683. As a student, Sherard began a friendship with Jacob Bobart the younger, which continued for the rest of Bobart's life. Bobart was instrumental in developing Sherard's botanical interests, which were augmented under the tutelage of Joseph Pitton de Tournefort in Paris, and of Paul Hermann and Herman Boerhaave in Leiden, in the latter part of the century.

On returning to England, Sherard became a travelling companion or tutor to various members of the aristocracy and landed gentry. In 1703, he was appointed to the prestigious and well-remunerated position of consul at Smyrna (modern İzmir, Turkey) for the Levant Company. Fourteen years later he returned to England as a wealthy man.

During the final years of his life, Sherard was based in London, and became absorbed by his *Pinax* project, in which he hoped to create a catalogue of the world's plant names. However, he continued to travel to the Continent; cultivate botanical relationships; acquire, maintain and renew friendships; and add to his herbarium. Sherard is buried in Eltham, the home village of his brother James Sherard. Linnaeus named the Eurasian plant genus *Sherardia* after William Sherard.

1

2

3

h

a

C. Mathews. Del & Sc.

Sherárdia arvensis Blue

Pub.d by W. Baxter Botanic Garden Oxford

5

rdia ⊙

Field madder (*Sherardia arvensis*), named in honour of William Sherard, from a hand-coloured engraving in William Baxter's *British Phaenogamous Botany* (1839). Bodleian Library, Sherardian Library of Plant Taxonomy, (4.1) BA2Dd, t.244.

Charles Giles Bridle Daubeny, FRS

(1795–1867),[48] a Gloucestershire-born chemist, geologist and botanist, won a scholarship to Magdalen College, Oxford, in 1810. Except for a brief spell in Edinburgh to study medicine, Daubeny remained at Magdalen for the rest of his life.

In 1822 Daubeny succeeded to the Aldrichian Chair of Chemistry, where his research focused on the chemical theory of volcanic action. In 1834 he was elected the fifth Sherardian Professor of Botany, and in 1840 to the first Sibthorpian Chair of Rural Economy. Daubeny capitalized on the clutch of academic chairs he held to investigate the links between chemistry, geology and botany. As had been done with botany two generations earlier, Daubeny fought to break the long-held idea that chemistry was a servant of medicine. Within the university, Daubeny was an advocate of educational reform. However, despite his advocacy of teaching, at least one student reported that he was reluctant to 'condescend to rudimentary teaching'.[49]

For his research on plant nutrition the German chemist Justus von Liebig characterized Daubeny as 'the zealous propagator of scientific principles in agriculture'.[50] Daubeny transformed the Botanic Garden estate largely at his own expense, and his work in plant physiology and fungal

biology briefly turned the Garden into a respectable place to be taught and in which to conduct botanical research in the mid-nineteenth century. His mind was open to new ideas, and he looked beyond Oxford through his travel, fieldwork and networks of correspondence. John Lindley commemorated Daubeny in the South African genus *Daubenya* because his 'interesting researches in Vegetable Chemistry have materially conduced to improve our knowledge of the physiology of plants'.[51]

Henry Borron Fielding (1805–1851)[52] was the only son of the successful owner of a Lancashire calico-printing company. A retiring man of indifferent health, Fielding used his substantial inheritance to pursue his botanical interests. In 1836 he began purchasing entire herbaria from collectors around Europe. He expanded his activities by subscribing to botanical fieldwork overseas, where he provided financial support for collectors in return for some of the specimens, and by purchasing herbarium specimens at auctions. After fifteen years, Fielding owned one of the best private herbaria in nineteenth-century Europe.

Fielding and his wife, Mary, took curation seriously, creating a plant collection known for the quality of its preservation, the order of its specimens and the breadth of its coverage. As the collection grew, at least one house move was prompted by fears for the safety of the specimens. Mary donated the herbarium and part of Fielding's library to the university in 1853, with enough money to employ a curator.

Fielding began to prepare notes on the more unusual specimens in his collection. The plan was to publish these, along with illustrations by Mary, as the *Sertum plantarum*. The Fieldings were helped in their task by the short-lived appointment, at the suggestion of the director of Kew, William Jackson Hooker, of the young Scottish botanist and tropical explorer George Gardner. Gardner was to give

Fielding's collection 'a more scientific management' and to identify 'the species of many of the specimens'.[53] After less than a month, Gardner accepted Hooker's offer to become director of the Peradeniya Botanic Garden in Ceylon (Sri Lanka). Understandably, Fielding was annoyed at losing the 'benefit of M^r. Gardner's scientific knowledge'.[54]

The *Sertum plantarum* became a friction point between the pair. Gardner felt that his contributions deserved equal status to those of Fielding. A mutual friend intervened to calm the furious Gardner. The grudging compromise was to add 'assisted by George Gardner' to the title page. Gardner accepted the situation but considered Fielding 'a man woefully deficient in moral integrity'.[55]

Specimen of *Qualea gardneriana*
from Fielding's herbarium, collected by ranch workers for the Scottish surgeon George Gardner in Brazil in 1839. Oxford University Herbaria.

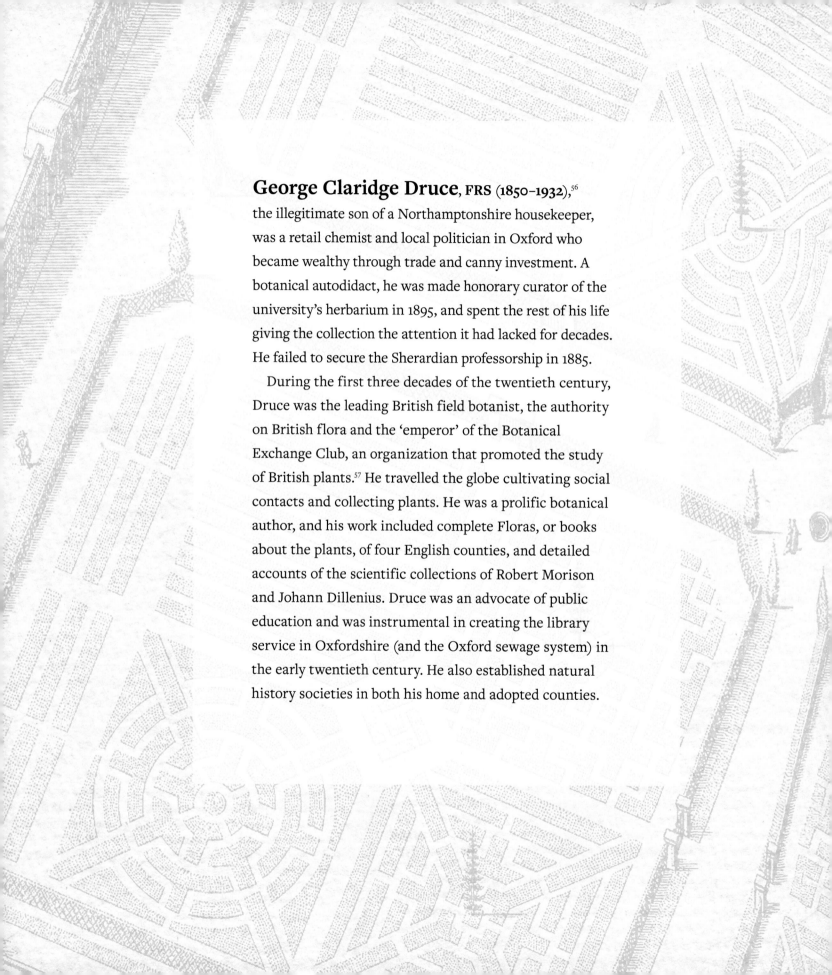

George Claridge Druce, FRS (1850–1932),[56] the illegitimate son of a Northamptonshire housekeeper, was a retail chemist and local politician in Oxford who became wealthy through trade and canny investment. A botanical autodidact, he was made honorary curator of the university's herbarium in 1895, and spent the rest of his life giving the collection the attention it had lacked for decades. He failed to secure the Sherardian professorship in 1885.

During the first three decades of the twentieth century, Druce was the leading British field botanist, the authority on British flora and the 'emperor' of the Botanical Exchange Club, an organization that promoted the study of British plants.[57] He travelled the globe cultivating social contacts and collecting plants. He was a prolific botanical author, and his work included complete Floras, or books about the plants, of four English counties, and detailed accounts of the scientific collections of Robert Morison and Johann Dillenius. Druce was an advocate of public education and was instrumental in creating the library service in Oxfordshire (and the Oxford sewage system) in the early twentieth century. He also established natural history societies in both his home and adopted counties.

Druce was a complex character who was exalted and reviled in equal measure. It was unusual for someone from his social background to make the inroads that he did into the British aristocracy, learned societies and academia. Druce bequeathed his vast herbarium and the bulk of his fortune to the university. He believed that he had provided the resources needed to look after his collections, and even to create a botanical institute. As we shall see, in this he was mistaken (see Chapter 7).

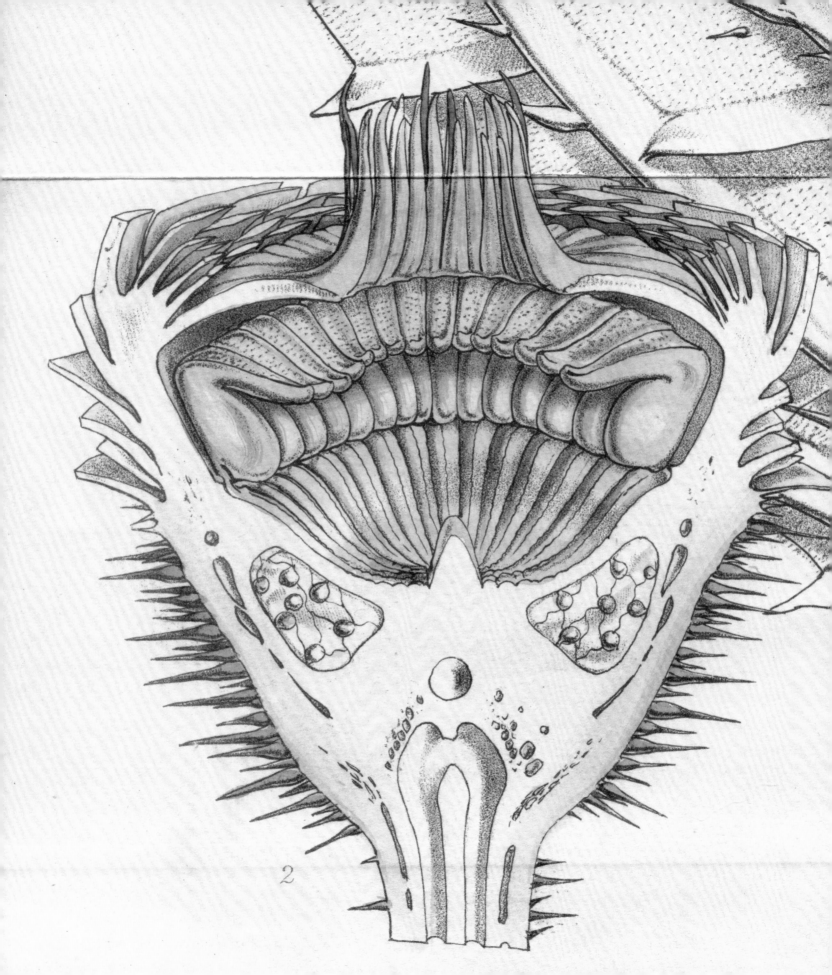

2

3 LEAF

Collectors and Collecting

Over the past four centuries, thousands of people have collected millions of plant samples to enhance our knowledge of plant biology at Oxford. Most of these samples were transitory: harvested from the Garden or the field, then discarded once specific data were collected. Only a tiny fraction of samples have been preserved as herbarium specimens and seeds, or as illustrations and photographs. This chapter focuses on the collectors who have contributed to the preserved objects in Oxford's botanical collections, rather than the collectors of transitory objects associated with plant sciences research.

Collectors must not be underestimated. Plant exploration is a serendipitous activity, involving plants that attract attention at specific times and in specific places.[1] The decisions collectors make in the field will determine the quality of a specimen and the data associated with it. Collecting is the earliest stage in the scientific enterprise and collectors' decisions and professionalism have repercussions for all subsequent uses of their work. Samples that are little better than scraps, with poor associated data, are hardly worth the expense of preservation or of data extraction and analysis.

The motivations of collectors who have contributed to the university's collections are manifold. For some, plants were merely another spoil of adventure, or a means to personal fame; others were motivated by a desire to know or to answer specific questions, or might have been inspired by an idea. Expeditions were financed through personal fortunes, the generosity of philanthropists, the university or the largesse of governments. Collectors travelled individually, as members of ad hoc expeditions or commissioned enterprises. For some collectors the rewards are recognition by the few during life or after death, but most are merely footnotes for having contributed a few data points to the research of others.

Hand-coloured lithograph of flower and leaf sections of the Victoria water lily (*Victoria amazonica*) (*detail; see p.77*)

Herbarium specimen of Victoria water lily (*Victoria amazonica*) cultivated at Chatsworth House, c.1850, by Joseph Paxton and given to Henry Fielding. Oxford University Herbaria, Paxton s.n.

Herbarium specimen of the state flower of Western Australia, Sturt's desert pea (*Swainsonia formosa*), collected by the English privateer William Dampier in Western Australia in August 1699. Oxford University Herbaria, Sher-0015-a.

THE VICTORIA WATER LILY

In the late 1810s the French botanist Aimé Bonpland, who had been the companion of Alexander von Humboldt during his Latin American explorations between 1799 and 1804, spotted the Victoria water lily (*Victoria amazonica*) floating on a small river near Corrientes in Argentina.[2] He 'well nigh precipitated himself off the raft into the river in his desire to secure specimens', and was, apparently, 'able to speak of little else for a whole month'.[3] In 1845, when he harvested the seeds of the water lily, which he eventually sent back to England packed in a ball of clay, the English plant collector Thomas Bridges stated: 'fain would I have plunged into the lake to procure specimens of the magnificent flowers and leaves; but knowing that the waters abounded in Alligators, I was deterred from doing so'.[4]

Each floating, dish-like leaf, which can be up to two metres in diameter, is supported by a lattice of prominent, spine-covered veins. Bridges found lakes covered with these immense leaves, vying for space with the flower buds which, raised above the surface of the water, were approximately the size of a child's head. The size of the plants presented him with a collecting challenge: only two leaves fitted into his canoe at a time, so numerous journeys were necessary to procure all the specimens of leaves, flowers and fruits he desired for European plant collections. Bridges ingeniously solved the problem of transporting his trophies over land by 'suspending them on long poles with small cord, tied to the stalks of the leaves and flowers'; the Amerindians who carried the poles were 'wondering all the while what would induce me to be at so much trouble to get at flowers'.[5]

The plants had been known from illustrations since the early nineteenth century, but, with the arrival of the Victoria water lily seed in England, competitive horticulture flourished. The race to get it to flower in cultivation was won by Joseph Paxton in 1849, in glasshouses at Chatsworth House, Derbyshire. When the water lily was shown to the public, Paxton the showman placed his daughter dressed in a fairy costume on one of its leaves.[6] The following year, he designed a simple rectangular glasshouse, based on the extraordinary ribbing of the leaves, for the water lily. The design eventually became the basis of the Crystal Palace at the Great Exhibition of 1851. In Oxford, Charles Daubeny succumbed to the water lily's charms, and had a tank and glasshouse constructed at the Botanic Garden to house it.

With the growth of field exploration, generations of Victorian naturalists observed the water lily's biology, especially its flowering. The Anglo-German explorer Robert Schomburgk found beetles in the flowers, while the French botanist Jules Planchon, who is credited with saving the French wine industry from the phylloxera bug in the nineteenth century, reported that temperatures inside the flowers were higher than those outside. Bridges observed that each spine-covered bud expanded in the evening to produce a pure white flower that was rose-pink by the morning, and that wild water lily populations had flowers at many different stages. The flowers also had a rich fragrance like a 'Pine-apple, afterwards to a Melon [*Cucumis*], and then to the *Cherimoya* [custard apple, *Annona cherimola*]'.[7] Connections between colour changes, beetles, temperature differences and fragrances were established in 1975 by *an* Oxford botany graduate, Ghillean Prance, who became director of the Royal Botanic Gardens Kew, and his Brazilian collaborator Jorge Arius.[8]

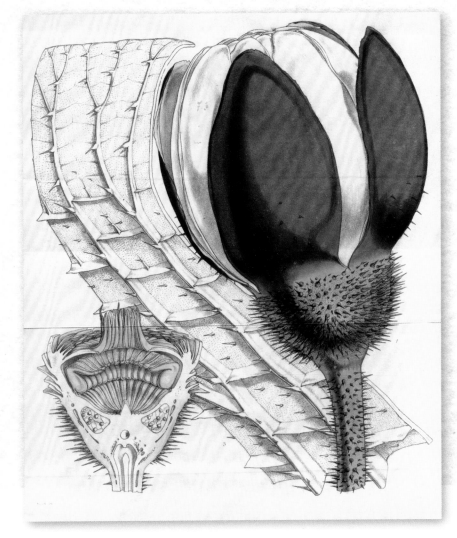

Hand-coloured lithograph of a bud and flower and leaf sections of the Victoria water lily (*Victoria amazonica*), by Walter Fitch, published in *Curtis's Botanical Magazine* (1847). Bodleian Library, Sherardian Library of Plant Taxonomy, t.4277.

Field skills

Labelled herbarium specimens, the gold standard of botanical documentation, are the physical evidence for a species' occurrence at a specific point in time and space. With specimens, botanists avoid the problems of translating literature records of vernacular names, variously applied scientific names or ambiguous images into scientific names across time and cultures. Even poor specimens are better than trying to interpret limited descriptions by poorly trained observers, using arbitrary technical language, especially when differences between species are subtle. High-quality specimen preparation, which is essential if the specimen is to be useful for research, is an apparently simple process but requires attention to detail: 'there are many ways of making an herbarium, but few ways of making a good one.'[9]

Plant collectors must be dedicated and adaptable. Instruction manuals provide advice on collection and preservation techniques, and expedition accounts and personal advice on the general conditions in a country or region. However, once they are in the field, plant collectors are alone and must respond to circumstances, modifying their activities in the light of experience and the conditions they find. The very best collectors are able to do this, and sometimes sustain the effort for years.

The Flemish anatomist Adriaan van den Spiegel published brief instructions for making herbarium specimens at the start of the seventeenth century, advising that samples – flowers, leaves and seeds – be placed between sheets of paper within the pages of a book. As the samples dried, they were to be checked daily, and weights were to be gradually applied to the book so that the specimens were kept flat.[10] About a century later, John Woodward made no fundamental changes to these instructions, but emphasized that it was essential to collect the correct material:

> As to *Plants* ... *four Samples of each kind* ... will be sufficient. Where the Plant is large, as in *Trees*, *Shrubs*, and the like, a *fair sprig*, about a *foot* in length, with the *Flower* on, ... but of the *lesser Plants*, such as *Sea-Weeds*, *Grasses*, *Mosses*, *Ferns*, &c. take up the *whole Plant*, root and all. Chuse all these Samples of Plants *when they are in prime*, I mean in *Flower*, *Head*, or *Seed*, if possible; And if the *lower* or *ground Leaves* of any Plant de *different* from the *upper leaves*, take two or three of them, and put them up along with the Sample.[11]

Herbarium specimen of white water lily (*Nymphaea alba*) collected by Peter Collinson, the gardener and friend of Johann Dillenius, from pools of water at Brockenhurst, Hampshire, in 1739. Oxford University Herbaria, Syn-368.3.

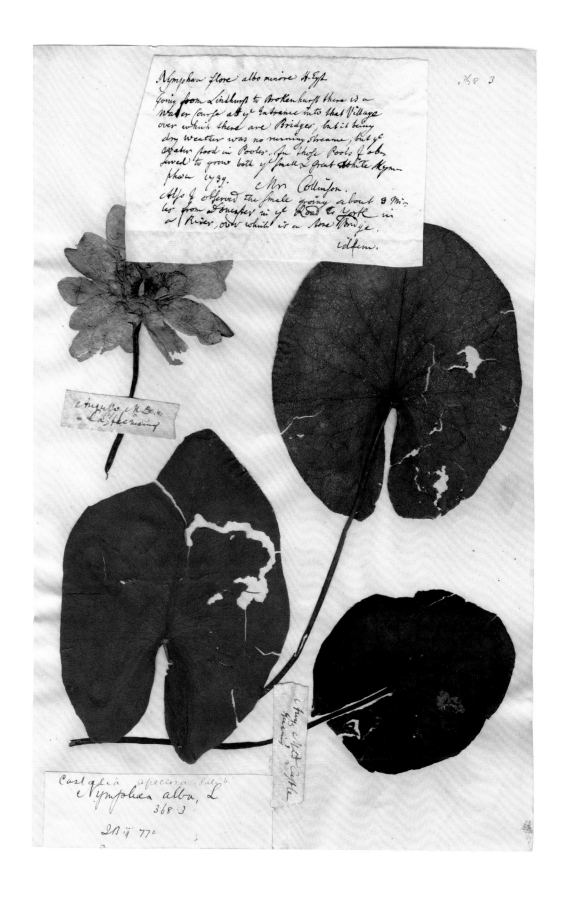

Nymphaea flore albo minore H. Eys

Going from Lindhurst to Brokenhurst there is a
Water Course at ye Entrance into that Village
over which there are Bridges, but it being
dry weather was no running Streame, but ye
Water stood in Pooles. In those Pools I ob=
served to grow both ye Small & great White Nym=
phaea 1739.   Mr Collinson.
Also I observed the Small going about 3 mi=
les from Doncaster in ye Road to York in
a River, over which is a Stone Bridge.
                              idem.

Castalea speciosa Salysb
  Nymphaea alba, L
       368·3

   2B ij 770

Vasculum, penknife, hand lens and map used by George Claridge Druce when collecting plants in the United Kingdom in the early twentieth century. Oxford University Herbaria, uncatalogued.

These were the methods adopted by plant collectors who added specimens to the Oxford herbarium. At the beginning of the nineteenth century, the fundamental principles remained unchanged, but the paraphernalia of plant collection had increased.[12]

When William Jackson Hooker, then director of Kew, released his cursory instructions for 'use of officers in Her Majesty's Navy and travellers in general', in 1849, he made sure that they were simple and practical: press in 'such a manner that their moisture may be quickly absorbed, the colours, so far as possible, preserved, and such a degree of pressure imparted that they may not curl in drying'. To do this it was necessary to have abundant paper 'of moderate folio size and rather absorbent quality'.[13] Dried specimens were then protected in paper folders and boxed for storage and transport. The task of drying absorbent paper is one of the many laborious operations collectors face when preparing specimens in the field. Such mechanical tasks in the physical collection and preservation of natural history objects could be delegated to the '*Hands of Servants*', although they were to be done in 'their spare and *leisure times*'.[14]

In the nineteenth century, the vasculum began to be recommended for fieldwork:

> if the specimens cannot be laid down as soon as gathered, they should be deposited in a tin box [vasculum], which indeed is essential to the botanist when travelling; there they will remain uninjured for a day and night, supposing the box to be well filled and securely closed to prevent evaporation.[15]

Initially, the vasculum was a rare, coveted object and, as well as being practical, it set apart the serious botanist, as the hand lens also did.[16]

Plant collectors were encouraged to collect the relevant information and even to use covert means of plant collection for 'countries which we are not permitted to explore'; for example, 'the curious traveller may obtain many rare plants, if he will examine the fodder that is brought down from the country, by the natives'.[17]

If natural history objects so carefully selected, collected and preserved in the field were to be of scientific use, they had to be safely returned to Britain. Collectors who were in the field for long periods would return specimens in batches to spread the risk of loss, maintain their income and create space for the collection of more specimens. Woodward was concerned with packing specimens so 'that the things be not *broken*, or *rifled* and *confounded* by the *Custom-house Officers* and *Searchers*'.[18]

A plant collector's field kit has changed over the past four centuries. Presses, paper, notebooks and rations are constants in the collection enterprise, with the addition of a lens for examining small objects by the beginning of the nineteenth century. Maps, a compass and clocks helped collectors to determine their position, while altitude could be measured using mercury-filled barometric tubes or by boiling water: all these

Pressed leaf of American lotus (*Nelumbo lutea*) collected by Mark Catesby in the Carolinas, colonial North America, and sent to William Sherard in 1722 accompanied by a field sketch of the flower (see p.89). Oxford University Herbaria, Sher-1090.

measurements can now be accomplished using the Global Positioning System (GPS). Historically, a collecting kit sometimes also included weapons and ingenious methods for scaling trees.[19]

Richard Richardson

In the early days of the Oxford collections, plant collectors had no need to venture beyond British shores to enhance their understanding of plant biology. Between 1658 and 1662 the English naturalist John Ray, who was based in Cambridgeshire, undertook a series of extended collecting expeditions into North Wales, the Peak District, the Welsh borders, Cumbria, Yorkshire, Lincolnshire and Scotland. Ray's explorations revealed the diversity and distribution of British plants, and provided encouragement for other botanists.[20]

One of these men was the wealthy Yorkshireman Richard Richardson. Richardson matriculated at Oxford in 1681, but within months was pursuing his medical education in Leiden, one of the centres of European botany in the seventeenth century.[21] By 1690, the newly trained Richardson had returned to Yorkshire with a deep interest in natural history, especially overlooked plants such as mosses and lichens. For the next fifty years, Richardson amassed a natural history library, supported the work of early eighteenth-century naturalists and travelled widely in England, Wales and Scotland collecting plants.[22] Yet, by the end of the eighteenth century, he was largely unknown, for he published little: he had decided that he did not want his name to appear in print.[23] However, the traces of his influence as a collector of books, plants and people can be found in his networks of correspondents. His Oxford network included Jacob Bobart the younger, Edward Lhwyd, William Sherard and Johann Dillenius.

In the late 1690s Lhwyd was exploring the natural history of North Wales and collecting living plants and herbarium specimens that he returned to Bobart in Oxford, among other correspondents. Richardson, who was one of these, joined Lhwyd on a journey into the Welsh mountains in 1700. Twenty-five years later, when Dillenius visited the area, Richardson commiserated with him over the tough living conditions: 'your accommodation at the Alehouse must be very mortifying, for I am sure you could meet with neither meat nor drink nor lodgings fit for any person but a native of the place.'[24] Richardson ignored Dillenius's unsubtle hint:

> If some rich Botanist, that hath no family and children, would
> build a house there, and buy some land to it, which might be
> done with a little money, it would be a very kinde invitation for
> Botanists to visit these strange places.[25]

Many of the specimens Richardson sent to Bobart, Sherard and
Dillenius are unusual for the period; they are accompanied by a
detailed locality and, frequently, ecological information. For example,
a specimen of hairy rock-cress (*Arabis hirsuta*) bears a label in
Richardson's distinctive hand: 'On the rocks nigh Malha[m] & Setle in
Craven pretty plentyfull but in abundance upon the Walls of Carlton
Hall & especially upon the Court Walls an abundance. Carlton is six
miles from Skipton in Craven.'[26] As specimens from other collectors
circulated among members of his networks, he added his own
comments to them. A specimen of hoary whitlow grass (*Draba incana*)
includes the comment that

> Mr Lhwyd never sawe this plant in flower, nor I in its native
> place where its very rare to be met with, he found it upon the
> mountaine Husvae the plants I met with grew upon the moist
> rocks of phynon vellon nigh Llanperis being planted in the
> garden it flowers & seeds in abundance so ... it is knowe very
> common here.[27]

Through his own collections from the wild and his exchanges with
other naturalists and horticulturalists, Richardson's garden at Bierley
Hall in west Yorkshire was reputed to be the best in the north of
England.[28] Ever hungry for new acquisitions, Richardson also used his
garden as a scientific collection to see whether the same species that
looked different in different places retained these differences when
grown side by side.[29]

Richardson's plant specimens, scattered through herbaria in Oxford
and at the Natural History Museum in London, provide us with data for
investigating plant distributions. His prolific correspondence gives us
insights into the practice of botany in eighteenth-century Britain and
the views of Oxford-based botanists, which would have been otherwise
lost. In a postscript to a letter dated 25 August 1736, Dillenius tells
Richardson that

A new botanist is arose in the north; a founder of a new method 'a staminibus et pistillis', whose name is Linnaeus ... He is a Swede, and hath travelled over Lapponia; hath a thorough insight and knowledge of botany; but am afraid his method wont hold. He came hither and stayed here about eight days.[30]

It was to Richardson that Dillenius expressed his concern that he might not get a position at Oxford and complained about James Sherard's behaviour following the death of his benefactor, William Sherard. Richardson's correspondence also reveals that he was Lhwyd's confidant, and the man with whom William Sherard shared his concerns over the university's treatment of Jacob Bobart the younger towards the end of his life.[31]

Thomas Shaw in North Africa

In 1720 Thomas Shaw, the son of a prosperous Westmorland cloth worker, arrived in Algiers as chaplain to a trading company called the English Factory, having completed his theological training at Queen's College, Oxford. Shaw's robust constitution and a 'mind rich in most kinds of human learning',[32] together with his light professional duties in Algiers, enabled him to become a respected explorer of North Africa and the eastern Mediterranean. Much of his exploration took place in the interior of what the British called Barbary (Algeria, Morocco and Libya), with forays into Tunisia, Egypt, Sinai and the Holy Land. After his return to England in 1733, he was elected a member of the Royal Society in 1734 and made principal of St Edmund Hall in 1740.

Despite collecting more than 600 plants, 140 of which he believed to be new species, Shaw dwelled little on how or where the specimens had been collected.[33] His 'Catalogue of some of the rarer plants of Barbary, Egypt and Arabia' was prepared with 'great Assistance from Mr. Professor Dillenius, whose Character in Botanical Learning, is so well known to the Publick'.[34] Importantly, Shaw deposited his specimens in the university's herbarium. He recognized explicitly: (1) the importance of physical specimens being permanently preserved in collections; (2) the possibility that his identifications were incorrect; and (3) that there were alternative interpretations of the plants he had collected.[35] Despite Shaw's foresight, and the knowledge that his specimens were in Oxford, little use has been made of the material. Some of Shaw's species were

Lichen specimens associated with Johann Dillenius' *Historia muscorum* (1741), mounted with their distinctive border of eighteenth-century wallpaper. Oxford University Herbaria, HM_68.

(a)   LXVIII.

73. Lichenoides tenue & molle, Agarici facie.   *The XXIV.*
thin and soft Agaric-like Lichenoides.

74. Lichenoides cartilagineum, scutellis fulvis pla- XXIV.
nis   *The cartilaginous Lichenoides, with larg*

B. Varietas
pulverulenta &
crispa.

XXIV.   75. Lichenoides imbricatum viridans, scutellis ba-
A. B.   diis.   *The greenish Lichenoides, with Chesnut
            colour'd Dishes.*

Lichenoides vulgare sinuosum, foliis & scutellis luteis Hist. Musc. p. 180. n.76.

cited by Carl Linnaeus in his *Species plantarum* (1753), while the French botanist René Desfontaines used about 40 per cent of Shaw's names in his *Flora Atlantica* (1798–99). Neither botanist appears to have examined the original specimens.

Shaw was primarily interested in geography, antiquities and local customs, although he described the published account of his explorations, which is prefaced with effusive thanks to George II and Queen Caroline, as 'an Essay towards restoring the antient Geography, and placing in a proper Light the Natural History of those Countries'. Aware that travellers' tales about 'Diet and Reception of the Traveller; the Hardships and Dangers to which he is exposed' would interest readers, he focuses in his preface on such 'Matter of too great Curiosity'. Shaw's approach to publication was praised as 'instructive and entertaining', avoiding 'the dullness of superfluous description, the cruel tediousness with which many a modern naturalist burdens his unhappy readers'.[36]

In the coastal towns of Barbary and the eastern Mediterranean, Shaw was entertained by other British trading companies with 'extraordinary Marks of Generosity and Friendship'. He benefited from their horses, their servants and, without commenting on it, the abhorrent practice of slavery. Inland, alive to the danger of bandits, Shaw stayed with local communities rather than have his team appear as 'Persons of Rank and Fortune, and consequently too rich and tempting a Booty to be suffered to escape'; his best protection was to 'dress in the 'Habit of the Country'. Two generations later, John Sibthorp had similar fears during his exploration of the eastern Mediterranean, even telling students about the early eighteenth-century French botanist Joseph Pitton de Tournefort: 'inflamed with an Ardour for Discovery he braved the greatest Difficulties & exposed Himself alone on the Pyrenees to the Insults & Maltreatments of yᵉ. Miquelets [a term used for Catalan bandits]'.[37]

Shaw repaid hospitality variously with gifts of a 'Knife, a Couple of Flints, or a small Quantity of English Gunpowder', or 'a Skean of Thread; a large Needle; or a Pair of Scissars'. His travelling choices sometimes meant that his small team had 'nothing to protect us from the Inclemency either of the Heat of the Day, or the Cold of the Night, unless we met with some accidental Grove of Trees, the Shelve of a Rock, or sometimes, by good Fortune, a Grotto'.

Although plant collecting can be a dangerous activity, generally Shaw travelled with minimal threats to his safety in North Africa. In the Holy

417 Myrrhis annua Luſitanica, ſemine villoſo,
Paſtinacæ ſativæ folio I.R.H. 315. Panax Sicu-
lum &c. Bocc. Rar. 1.

BARBARY
Dr SHAW circa 1700

Land, however, he was constantly aware of bandits. In 1722, travelling between Ramah (probably modern Er-Ram) and Jerusalem, 'four Bands of Turkish Soldiers ... were not able, or durst not at least protect us, against the repeated Insults and Ravages of the Arabs'. Besides dangers from other people, fleas, lice and the 'Scorpion, viper, or Venemous-Spider', Shaw feared being trampled by the young of livestock tethered around the places where he stayed. When forced to travel at night they heard 'Lyons roaring after their Prey; the Leopards, Hyaenas, and a Variety of other ravenous Creatures, calling to and answering each other'.

Shaw highlighted the deprivations and dangers of exploration:

> from Kairo to Mount Sinai, the Heavens were every Night our only Covering; the Sand, spread over with a Carpet, was our Bed; and a Change of Raiment, made up into a Bundle, our Pillow. Our Camels (for Horses or Mules require too much Water to he employed in these Deserts) were made to lye round us in a Circle, with their Faces looking from us, and their respective Loads and Saddles placed behind them. In this Situation, they served us as so many Guards, being watchful Animals, and awaking with the Least Noise.

When times were good the animals had 'Barley, with a few Beans intermixed; or else the Flour of one or other of them, made into Balls', while Shaw dined on 'Wheat-Flour, Biscuit, Honey, Oyl, Vinegar, Olives, Lentils, potted Flesh, and such Things as would keep, during two Months'. When times were tough Shaw's team gathered 'Stubble, Grass, Boughs of Trees' for the animals before they were forced to eat their only provisions, the 'Fragments, of some former Meal'. Camel dung, which 'catches Fire like Touchwood, and burns as bright as Charcoal', was their fuel for cooking.

Mark Catesby in the Americas

North America offered the prospect of great riches to natural historians in the eighteenth century.[38] Hans Sloane explored Jamaica in the 1680s, and exchanged specimens with William Sherard. In the 1710s Richardson and Sherard sent the botanist Thomas More to collect in New England, but 'very few profits to science accrued ... the scanty returns clearly having been the consequence of [his] carelessness'.[39] What such gentlemen needed was a skilled, dedicated and reliable collector. They found their man in the English naturalist Mark Catesby.

As he explored the southern North American colonies, Catesby collected all sorts of objects for sponsors with multifarious desires and little understanding of the conditions under which he was working. If he was to succeed, he had to manage his sponsors' expectations, preserve animals and plants, collect living plants, protect his collections from pests, fungi and damp, and record what he observed.[40]

For four years, between 1722 and 1726, Catesby travelled through the Carolinas into Georgia and Florida and finally the Bahamas. However, most of the hundreds of extant herbarium specimens he collected lack even the minimal details we would expect to see on modern specimens. Catesby quickly realized that 'to collect everything is impossible but with many years application',[41] even though he had the advantage of living for protracted periods in the areas he explored. His approach to collecting was 'never to be twice at one place in the same season',[42] a strategy driven as much by expediency as by biology, for his sponsors wanted botanical novelties. Collecting seed and botanical specimens in the field was hard work and involved travelling great distances, sometimes in inclement weather, over hostile terrain. He faced the constant dangers of getting lost, encountering wild animals or meeting hostile indigenous peoples: 'the fear of ... meeting with Indians as five of us hapned to doe as we were out a Buffello hunting, tho' they hapned to be those we were at peace with they were about 60'.[43]

Once samples had been collected, the work was not over, for the samples had to be labelled, preserved, packed and transported. In addition, Catesby was also making drawings in the field. As he started to deliver on his promises, his work became harder. As more sponsors started to support him, he had to haul around more baggage, which would have included specimens in various stages of preparation, paper for drying plants and drawing gear, not to mention his food and clothes.

Some specimens proved a challenge to preserve, even for a determined collector such as Catesby. For example, associated with a leaf specimen of American lotus (*Nelumbo lutea*) sent to Sherard is a sketch of its flower that Catesby made in the field, together with the accompanying note, 'The flowr I could not preserve so have sent this scetch.'[44]

Little wonder, therefore, that Catesby resorted to the practices of the time and any resources available to him, including Native American

following pages **Mark Catesby's field sketch of the American lotus (*Nelumbo lutea*)** sent to William Sherard in 1722, from the Carolinas, colonial North America, accompanied by pressed leaves (see p.81) and a note about the plant's identity and ecology. Oxford University Herbaria, Sher-1090a.

guides and porters, and willing friends and acquaintances. It is shocking to discover that he even planned to purchase a slave, after petitioning William Sherard for the funds.[45] Catesby soon discovered that patronage comes with the burden of satisfying the competing demands of sponsors. Charles Dubois proved difficult to please: 'the discontent of Mr Du-Bois and the trouble he gives my Friends in receiving his Subscription is Such that I had rather be without it, I doubt Not that I have Suffered by his Complaints.'[46]

Having sent thousands of herbarium specimens and living plants, sometimes packed in precise and ingenious manners, to his sponsors, Catesby returned to England in 1726. His specimens, divided between Sherard, Sloane and Dubois, are preserved in the Oxford University Herbaria and at the Natural History Museum, London.[47] Specimens in the Sloane herbarium have been widely used for research. Those in the Sherard and Dubois herbaria have been overlooked, except by a few determined botanists such as the early nineteenth-century German American botanist Friedrich Traugott Pursh, who was delighted to find Catesby's specimens in Sherard's herbarium.

John Sibthorp and the *Flora Graeca*

John Sibthorp, the third Sherardian Professor of Botany, organized perhaps the most famous botanical expedition directly associated with the University of Oxford.[48] Unlike Mark Catesby, Sibthorp had complete freedom to explore his chosen area, the eastern Mediterranean. Independent wealth meant that he was free from the burden of satisfying the caprices of sponsors. Sibthorp also had a single-minded interest in botany (when in the field), an academic position that allowed him the time and freedom to undertake extensive fieldwork, access to the world's finest botanical collections and immense confidence in his own abilities. Such advantages were granted to few contributors to Oxford's botanical collections.

The botany of the eastern Mediterranean was not unknown before Sibthorp. Theophrastus and Dioscorides had written their classical works in the region. At the start of the eighteenth century the French monarchy sponsored an expedition led by Joseph Pitton de Tournefort, and in 1761 the Danish monarchy funded an expedition led by Carsten Niebuhr to the region. There are duplicates of Tournefort's specimens in the Sherard herbarium.

Watercolour of a rare Cypriot butterwort (*Pinguicula crystallina*) by Ferdinand Bauer, based on field sketches he made during his journey in the eastern Mediterranean with John Sibthorp, and completed in Oxford between 1788 and 1792. Bodleian Library, Sherardian Library of Plant Taxonomy, MS Sherard 244. f.53

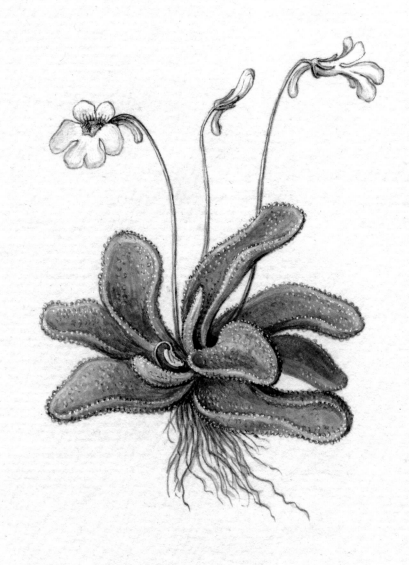

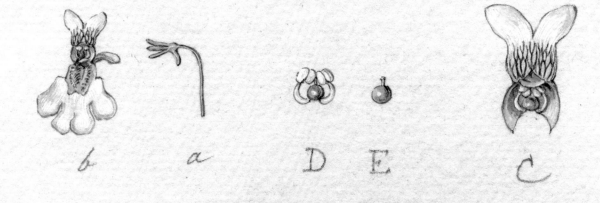

b   *a*   D   E   C

In preparing for his journey, Sibthorp picked two men who would become essential for his legacy; without them, he would probably have remained a botanical footnote. The two men were the Austrian botanical artist Ferdinand Bauer and the English mine owner and gentleman Hellenophile John Hawkins. Sibthorp was explicit about his motivation for undertaking the expedition:

> this is the part of my Voyage on which I form my greatest Expectations from whence I hope will flow a future Source of Fame, that tho' it does not place me in the Rank of Tournefort will give me the same Place with Hasselquist & Buxbaum without sharing the fate of one or the other. Many Discoveries are still to be made for the Natural History of these Countries if not imperfectly known. the figures of Buxbaum are ill executed they are no more reconnoissable. The Superiority of my Draughtsman will fence me what I shall publish from risking a similar Fate & will entitle me to a Place in the Petersburg or in our Academy with just Pretensions.[49]

Planning does not appear to have been one of Sibthorp's strong points. Bauer complains to Hawkins that 'I think Dr Sibthorp will beging His tour quite in the old style never come to a determination before the last day, then at once all in haste which is most unblessant'.[50] Planning is essential when plague, bandits and pirates are prevalent, although Sibthorp's plans had to be flexible, because such issues would become known only as they travelled. Money and contacts in the Ottoman empire were not a problem; he was, after all, a professor at Oxford, a wealthy man and a member of a social elite. However, there were things that, even with the best planning, Sibthorp could not control – the short flowering season of the plants and the weather.

Sibthorp was away from Oxford for over three years; he left his hometown in the late summer of 1784 and returned in early December 1787. During this journey, he spent extensive periods in Vienna, where he studied the Western world's oldest botanical manuscript and engaged Bauer. The journey took him from England through the present-day Netherlands, Germany, the Czech Republic, Austria, Slovenia, Italy, Greece, Turkey and Cyprus. Along the way, Sibthorp and his colleagues collected vast numbers of plants, animals and geological specimens, and

Bauer made sketches that would eventually become some of the finest botanical watercolours ever produced.

In Vienna, Sibthorp worked hard at gleaning what knowledge he could about the lands through which he planned to travel. He recognized the exceptional talents of his travelling companion Bauer: 'superior to any Artist I have yet seen' and 'my Painter in each part of Natural History is Princeps pictorum he joins to the Taste of the Painter, the Knowledge of a Naturalist & Animal, Plant & Fossil touched by his Hand shew the Master'.[51] However, over the years they worked together, the relationship between these two talented men soured.

Throughout their journeys Sibthorp collected plants and made notes, while Bauer sketched the plants and animals, recording colours using a numbering system. Botanical fieldwork is exciting, sometimes dramatic, but often stressful: and it is the completion of routine tasks that makes it successful. The most essential are collecting the specimens, pressing them and ensuring that they are properly dried. Once dried, specimens must be labelled and kept free from damp and pests. Sibthorp appears to have spent his mornings collecting plants with Bauer, and his afternoons describing them while Bauer sketched. However, Sibthorp did not label his specimens, trusting to his memory, which caused major problems for those responsible for piecing together the *Flora Graeca* after his premature death.

Sibthorp and Bauer left Vienna for Trieste on 6 March 1786 with a mixture of excitement and apprehension. At Trieste they crossed to Venice and then travelled through the cities of the Apennine (Italian) Peninsula to Naples. They left Naples in late May or early June and sailed through the Strait of Messina into a world that was poorly known botanically compared to the one through which they had passed.

By the end of June, they were in Crete. Their first excursion on Crete was to the Akrotiri Peninsula, considered by Tournefort to be one of the best places on the island for botanizing. Sibthorp enthused:

> we gathered the Ebeny of Crete (*Ebenus criticus* [modern name *Ebenus cretica*]), the Tree Pink (*Dianthus arboreus* [modern name *Dianthus juniperinus* ssp. *bauhinorum*]) the white Fleabane (*Conyza candida* [modern name *Inula candida*]), the Immortal of the East (*Gnaphalium orientale* [modern name *Helichrysum orientale*]) while the soft cotonny Dictammy carpeted its Sides,

among the Rocks we found many other curious Plants which the licentious Goats & the burning Sun had spared us ... A Plant that pleased me above all the rest was *Stahelina arborea* [modern name *Staehelina arborea*] of which we brought of a Tree covered with its Flowers & shining with its silvered Leaves.[52]

In the Sfaccia mountains, inhabited 'by Banditti most formidable & most treacherous',[53] Sibthorp used serendipity and guile. When he was consulted by the 'Charllatan' bashaw of Canea for a stomach disorder, Sibthorp claimed that he

was in Search of medicinal Plants. & that those which were best adapted to his Disorder grew on the Mountains of Sfaccia but where it was imprudent for me to go without his Protection & Guards which I requested Him to furnish me with.[54]

The guard forthcoming, the party explored the mountains, with Sibthorp stating, 'I now botanize always with loaded Pistols in my Pockets.'

After leaving Crete, Sibthorp and Bauer passed through Attica and the Aegean Islands before arriving in western Anatolia, on 1 August. The journey to Istanbul was probably made along the traditional caravan route, through Izmir, where Sherard had been ambassador.

Having followed almost exactly in Tournefort's footsteps, Sibthorp and Bauer reached the political heart of the Ottoman empire, Istanbul, in early September. They overwintered in the city. Having met up with Hawkins and Ninian Imrie, a British military engineer, the expanded expedition left by boat on 13 March 1787. They passed through the Dardanelles, skirting the Anatolian coastline. Near Akyar Burnu on the Asiatic mainland, one of the botanical prizes of the entire expedition was won by Imrie, 'who was by far the most agile of our party [and] after much labour and difficulty reach the summit of the nearest mountain & brought back with him a new and elegant species of *Fritillaria*'. This plant, named *Fritillaria Emereii* in honour of Imrie in Sibthorp's notes, had never been collected before, and is known only from extreme south-western Anatolia. Bauer made a sketch and eventually produced a magnificent watercolour, which was used in the formal description of the species, *Fritillaria sibthorpiana*, eventually named, rather unjustly, after Sibthorp rather than Imrie. It was more than 180 years before this

Entries from John Sibthorp's manuscript expedition diary for 5–8 August 1786, made during his exploration of the eastern Mediterranean. Bodleian Library, Sherardian Library of Plant Taxonomy, MS. Sherard 215, f.87r.

August 5.

Walked out on the shore to the North of the town the Rocks composed of serpentine with veins of Asbest & Soapstone intermixed. on the sample we collected some magnetic Iron ore with beautiful Chrystallisation. The Houses of Negro-pont have a mean appearance are mostly ill built & inhabited by Turks. The Greeks are here more oppressed than in the other Greek Islands & the Turks are said to have a bad Cha-racter - tho' we were assured by the French Consul that we might travel the Island in the greatest security. The shore of Boetia on the contrary is said to be dangerous as it was at present infested by pyrates. Aphrodite returning with the Firman of the Bashaw we set sail in the afternoon - the wind NE. we made different Tacks - at midnight the wind dying away we anchored under the lee of a Rock in the Island of Negropont.

August 6th.
we set sail at 7 in the Morn.

at 6 on the same Point of Negro... light continued to the westward of...

August 7th.
At two in the morn... the Grecian shore two draws a light ... a light breeze ... the north we began down the ... we were abreast of draws. We an...diately after we of Volo. - at two... went into a small... the North of the... Thessalian shore we put to again.

August 8th.

Early in the Morn at a small dist... Pallene a Prom... - during the...

above **Herbarium specimen of *Fritillaria sibthorpiana*** collected by the English mining engineer Ninian Imrie from mainland Turkey, opposite Rhodes, in 1787. Oxford University Herbaria, Sib-0790.

opposite **Field sketch of *Fritillaria sibthorpiana*** made by Ferdinand Bauer in 1787. Bodleian Library, Sherardian Library of Plant Taxonomy, MS. Sherard 247(6), f.21r.

horticultural novelty was collected again by a pair of Scandinavian botanists, Per Wendelbo and Hans Runemark.[55]

The expedition arrived in Cyprus in late March. Since they were staying on the island for about two months, they could collect much of the spring flora. The Cypriot flora had been poorly known before Sibthorp arrived, as the island had not been visited by Tournefort. When Sibthorp and Bauer left Cyprus, they had amassed collections of immense scientific value. They had listed or collected some 600 species on the island, including Cypriot endemics such as *Arabis purpurea*, *Onosma fruticosum* and *Silene laevigata*.

After leaving Cyprus, the expedition island-hopped across the Aegean when not becalmed. This must have been an immensely frustrating time for Sibthorp who was aware that the flowering season in the Mediterranean was short and that the more time spent at sea, the less time there was to collect new plants. By 19 June 1787 they had landed on the Greek mainland. A few days later, a visit to Mount Hymettos proved Sibthorp's point: 'parched & burnt up by the Sun & the few [plants] that remained were cropt by the Goats'.[56] He decided to focus on higher altitudes and the mountains, and thence followed excursions to Parnassus, Thessaloniki and Patras. A leisurely return to England through the Mediterranean saw Sibthorp and Bauer arrive in Bristol on 5 December 1787. A week later, Sibthorp, together with his 'Flora Graecia with Beasts Birds, & Fishes'[57] and artist, was in Oxford. They had returned with thousands of herbarium specimens and hundreds of pages of field sketches and notes. These materials would occupy Sibthorp for the rest of his life.

There were discussions about a second journey to the Ottoman empire between Sibthorp and Hawkins in 1789. Hawkins was very keen that Bauer should accompany them again, but Bauer was not interested, his relationship with Sibthorp having broken down irrevocably. Sibthorp had other problems: his family, particularly his half-brother and father, had reservations about his going on another European tour. One grand tour in a lifetime might be educational; a second would be avoiding his responsibilities as a gentleman and landowner.

After years of procrastination, Sibthorp finally set out for Istanbul in March 1794 at a time when the European political landscape was in great flux. The Terror was coming to an end in France, and Europe was in the throes of one its periodic internecine conflagrations. So Sibthorp chose

to travel to Istanbul via present-day Hungary, Romania and Bulgaria, covering approximately 2,800 kilometres in sixty days.

Sibthorp, never a man with a robust constitution, arrived seriously ill. His illness affected the rest of his journey as he explored further parts of Turkey and travelled to mainland Greece across the Aegean, where Zakynthos and Morea were his foci. He departed for England on 1 May 1795 and arrived in Oxford in early October 1795 a wrecked man.

On his first journey, Sibthorp had returned to Oxford with his specimens. On the second, he left the zoological material in Zakynthos to be forwarded to Oxford, and took his botanical specimens and notes with him. However, when he became seriously ill on the journey, he sent the botanical specimens back to Zakynthos for the British vice consul Spiridon Foresti to forward to England. With admirable attention to the details of packing, Foresti found safe passage for the zoological and botanical material on a ship sailing under a neutral flag. The material arrived in Oxford, but Sibthorp never saw it, for he died in Bath in February 1796.

George Claridge Druce in Oxfordshire

Long before the Garden's foundation, dons, physicians and students in the university had explored Oxford and Oxfordshire for plants. However, from the late 1870s until his death in 1932, the honorary curator of the university's herbarium, George Claridge Druce, dramatically changed how the county's plants were collected, and with this how much was known about them.

Before Druce, collectors were selective both in where they went in the county and in what and when they collected. In contrast, Druce roamed the county throughout the year, collecting anything he found of interest.[58] Furthermore, his collecting instincts were not contained by his adopted county. He collected throughout the British Isles, especially in his home county of Northamptonshire, and in Berkshire and Buckinghamshire.[59] By the time he died, Druce had collected more than 50,000 specimens, of which at least 10,000 were from Oxfordshire.

Druce was honorary secretary of the Botanical Exchange Club (later the Botanical Society of the British Isles), whose traditional members were specifically interested in the collection and exchange of rare and unusual herbarium specimens of British plants.[60] He believed that physical specimens should be prized as evidence for a plant's occurrence

Watercolour of *Fritillaria sibthorpiana* by Ferdinand Bauer, based on field sketches (see p.99) he made during his journey in the eastern Mediterranean with John Sibthorp, and completed in Oxford between 1788 and 1792. Bodleian Library, Sherardian Library of Plant Taxonomy, MS. Sherard 245, f.79r.

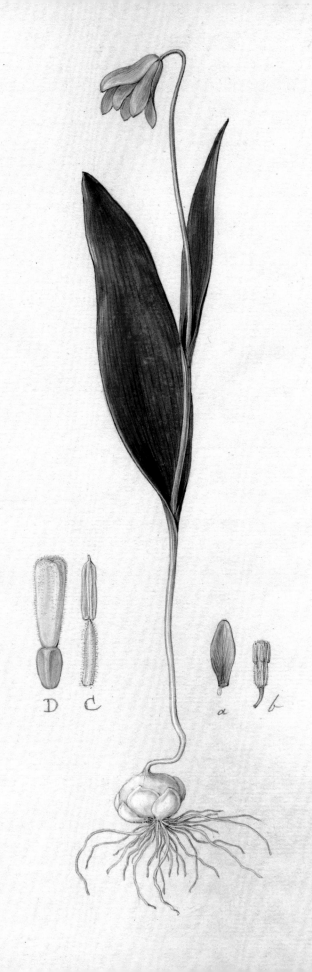

D   C   a   b

at a specific place, but many of his own collections rarely provide more detail than a village name and the year of collection.

Druce was interested in finding species in new localities and in discovering new variants of known species. He enjoyed being the first to discover a new plant. Working within the taxonomic climate of European botany in the early twentieth century, he gave hundreds of specimens in his collections the names of new variants; most of these are not formally recognized today, as they are regarded as usual variations within species. Nevertheless Druce's knowledge of minor variants was highly valued by William Bateson, one of the founders of the science of genetics and director of the John Innes Horticultural Institution between 1910 and 1926.[61]

Druce's indiscriminate approach to plant collection was notorious among his peers, even in a climate where species conservation was not a major concern.[62] In 1896 he gathered 'hundreds of specimens' of the rare grass *Bromus interruptus* within a 'few yards' of an arable field near Uxbridge.[63] His first reaction to hearing about a rare plant in the British Isles was to go and collect it.[64] A collecting companion who witnessed Druce harvesting armfuls of rare orchids in Cambridgeshire was reported to have required forcible restraint to prevent him hitting the old botanical reaper.[65]

Druce paid lip-service to plant conservation.[66] He was protective of the rarities he had discovered, but readily supplied specimens of very rare plants, roots and all, to other collectors and botanical artists. However, he was careful about whom he showed the plants growing *in situ*, perhaps fearing that others would behave in a similar manner to himself.[67]

Attitudes to Druce are often polarized; this was true during his lifetime and remains so today. Whether people loved or loathed him, however, they agreed that his knowledge of British flora of the first three decades of the twentieth century was unparalleled. Druce gained this knowledge, and the grudging respect of his peers, by tramping across Britain collecting plants and by trawling herbaria, libraries and archives for records of British plants. When the Cambridge professor of botany Arthur Tansley needed a guide to show British plants to delegates to the 1911 International Phytogeographical Excursion, and someone to sponsor the event, the wealthy Druce was an obvious choice, despite Tansley's reservations about the man.[68] Ironically, when Tansley became the tenth Sherardian professor in 1927, he became Druce's superior, although Druce's influence in British botany was diminishing by then.

An amateur botanist who, despite his best efforts, was always on the fringes of the university, Druce collected during a period when British attitudes towards botany were beginning to change. He had established himself when the tradition was bagging, tagging and cataloguing plants, but could see a future where plant conservation would be emphasized and the experimental approaches of biochemisty, ecology, genetics and physiology be brought to bear on the study of plants. The vast collections he bequeathed to a reluctant university remain largely unexplored for the contributions they could make to the study of plants in pre-1930s Britain.

Collecting living plants

Living plants provide multiple possibilities for botanical research and, in the modern argot, public engagement. At all stages in its history, the Garden has been stocked through plants being transported for what may be long distances as seeds or as rooted specimens.[69]

Bobart the younger, as supplier of rarities to the gentry, recognized the challenges of transporting living plants. In March 1694 he visited the celebrated gardens of the Duchess of Beaufort at Badminton in Gloucestershire. He was keen for her to purchase some of the plants he was raising: 'I send now a packet of such seeds as to me seem hopefull' and 'I send allsoe Madam a note of such good plants as I do not remember to have seen in yr Graces plantations.' However, the safe transport of the growing plants was the duchess's responsibility:

> The Plants herein mention'd, both in pots and wthout pots, may safely be handled, transplanted and carryed, about a fortnight hence; and if it may be consist wth yr Grace's pleasure to use any of them, it appears to me, the best way to send a man and horse of yr owne choosing, rather than commit them to the carelessness of a publick Carrier.[70]

Catesby faced similar challenges with the transport of living plants, and was forced to find novel solutions, including the use of gourds packed with damp sand. Even when seeds are safely in a gardener's hands, time, patience and luck are needed for successful germination.

Whereas a handful of seeds can be dropped in the pocket, growing plants must be cosseted in hostile environments such as a ship at sea.

A solution was found in 1823 when Nathaniel Bagshaw Ward, an amateur naturalist and general practitioner from the East End of London, made a breakthrough with the Wardian case, a closed, glazed box that protects growing plants from unfavourable conditions.[71] Before the Wardian case, approximately 99 per cent of plants imported to the United Kingdom from China were lost; after it was introduced, losses were reduced to 14 per cent. This simple technology was soon adopted worldwide, and became the standard way in which growing plants were moved around the world until after the Second World War. Ward also saw his cases in terms of 'their application to the relief of the *physical* and *moral* wants of densely crowded populations in the large cities'. As they became the terrariums of Victorian and Edwardian households, Wardian cases enabled many to glimpse the tropics. In the early 1920s two Wardian cases were being used in the Botanic Garden for a collection of filmy ferns, perhaps in a manner more akin to that envisaged by Ward rather than as an imperial technology.[72]

Today, high-speed transport has reduced the difficulties of moving living plants across the globe, but those of collecting living plants remain. In recent times, the Garden has been involved in few collection expeditions that have returned with living plants. In 1978 the unusual insectivorous plant *Heliamphora nutans* was collected on Mount Roraima in Guyana and propagated by the Garden's superintendent, Kenneth Burras. Seed-collecting expeditions undertaken by Ben Jones, the current curator of the Arboretum, and his Japanese collaborators have introduced Japanese plants of known wild provenance to the Garden and Arboretum.

Modern collecting

Over the past forty years, collecting for the university's collections has focused on specimens directly associated with research projects rather than, as in previous centuries, on the speculative accumulation of specimens. Growing the collection in this way is not a major motivation for collectors. The emphasis has been on forestry species, collaborative specimen collection expeditions and the effective involvement of people from the places where the specimens are collected.

In the early twentieth century, imperial, and then Commonwealth, forest officers swelled the shelves of the Herbaria through their interest in the application of taxonomy and ecology to the effective management of tropical forests, especially in Africa and South-East Asia.[73] Attentive curators and curatorial assistants, who often had extensive practical experience of collecting in the tropics, realized the importance of ensuring that specimens were available to researchers.

From the late 1920s, Joseph Burtt Davy and Arthur Clague Hoyle set the standard for combining collection care and practical fieldwork. Hoyle undertook extensive fieldwork in eastern and southern Africa, specializing in the ecologically and economically important tree legume genus *Brachystegia*. In the 1940s John Patrick Micklethwait Brenan worked in the herbarium and collected plants in East Africa. By the end of the decade, Brenan was at Kew, where he eventually become director in the mid-1970s.

Between the 1950s and the early 1990s, research by Frank White and David Mabberley, and their students and colleagues, focused on Africa and South-East Asia and the plant families containing the ebonies and mahoganies. These investigations involved extended periods of fieldwork across Africa, South-East Asia, Central America and Polynesia. Mabberley eventually became keeper of the Herbarium, Library, Art and Archives at the Royal Botanic Gardens at Kew.

From the 1980s to the early 2000s, the Forest Genetic Group, which was formed within the Forestry Institute, investigated the genetic resources of tropical trees. The methods used were the collection of seeds and herbarium specimens, together with the establishment of living provenance trials to compare the growth and production characteristics of trees. With funding from the British government's overseas development budgets, collectors worked with institutions across Central America and Africa to collect seeds from trees with multiple uses, especially legumes. Brian Styles collected pines throughout Central America, while researchers such as Colin Hughes and Duncan Macqueen made collections of tree legumes in the Americas. Richard Barnes and Chris Fagg searched for acacias in southern and eastern Africa and South Africa.

Today, a typical pattern of collection is based on geographically restricted trips of limited duration, by collectors early in their careers. A small number of collectors have made disproportionately large

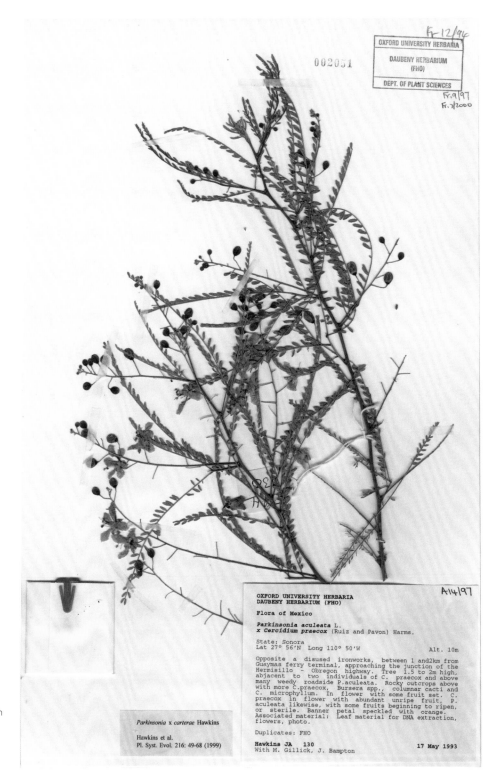

Herbarium specimen of the woody legume *Parkinsonia* x *carterae* described by Julie Hawkins in 1999, which she collected with colleagues in Mexico in 1993. Oxford University Herbaria, FHO_00164752P.

002051

Parkinsonia x *carterae* Hawkins

Hawkins et al.
Pl. Syst. Evol. 216: 49-68 (1999)

OXFORD UNIVERSITY HERBARIA
DAUBENY HERBARIUM (FHO)

Flora of Mexico

Parkinsonia aculeata L.
x *Cercidium praecox* (Ruiz and Pavon) Harms.

State: Sonora
Lat 27° 56'N Long 110° 50'W   Alt. 10m

Opposite a disused ironworks, between 1 and 2km from
Guaymas ferry terminal, approaching the junction of the
Hermisillo - Obregon highway. Tree 1.5 to 2m high,
abjacent to two individuals of C. praecox and above
many weedy roadside P.aculeata. Rocky outcrops above
with more C.praecox, Bursera spp., columnar cacti and
C. microphyllum. In flower with some fruit set. C.
praecox in flower with abundant unripe fruit. P.
aculeata likewise, with some fruits beginning to ripen,
or sterile. Banner petal speckled with orange.
Associated material: Leaf material for DNA extraction,
flowers, photo.

Duplicates: FHO

Hawkins JA 130   **17 May 1993**
With M. Gillick, J. Bampton

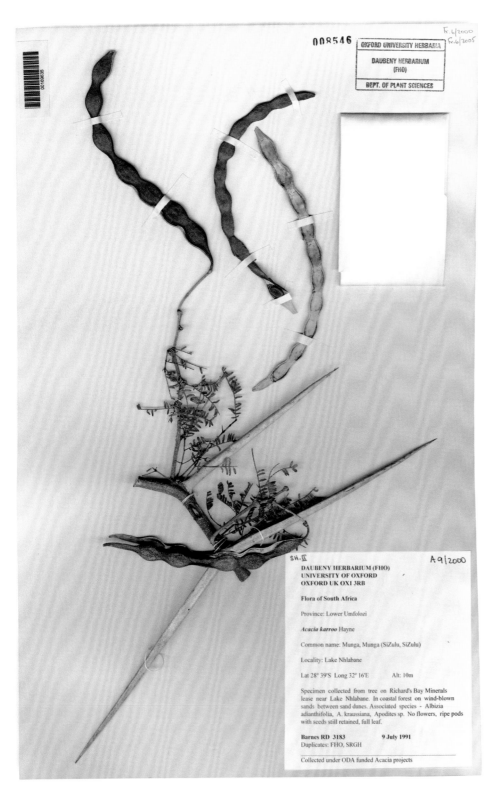

008546

F. 6/2000
Fr.6/2005

SH. II

A 9/2000

DAUBENY HERBARIUM (FHO)
UNIVERSITY OF OXFORD
OXFORD UK OX1 3RB

Flora of South Africa

Province: Lower Umfolozi

Acacia karroo Hayne

Common name: Munga, Munga (SiZulu, SiZulu)

Locality: Lake Nhlabane

Lat 28° 39'S Long 32° 16'E   Alt: 10m

Specimen collected from tree on Richard's Bay Minerals
lease near Lake Nhlabane. In coastal forest on wind-blown
sands between sand dunes. Associated species - Albizia
adianthifolia, A. kraussiana, Apodites sp. No flowers, ripe pods
with seeds still retained, full leaf.

Barnes RD 3183   **9 July 1991**
Duplicates: FHO, SRGH

Collected under ODA funded Acacia projects

Herbarium specimen of *Acacia karroo* collected by Richard Barnes in South Africa in 1991 as part of seed collection programmes to study the use of African multipurpose legumes. Oxford University Herbaria, FHO_00169635U.

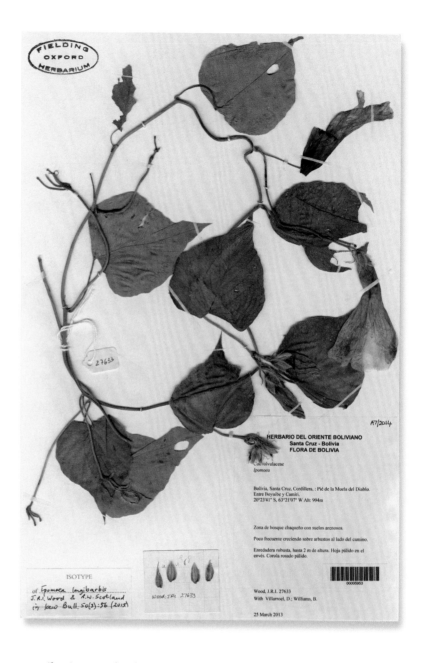

Type specimen of *Ipomoea longibarbis*, a member of the sweet potato family, collected in Bolivia by John Wood and colleagues in 2013, and formally described as a new species by John Wood and Robert Scotland two years later. Oxford University Herbaria, OXF_00005953.

contributions to the discovery of new plant species. Since the 2000s one of these, John Wood, has been associated with Oxford through his collaborations with Robert Scotland.[74] One of Wood's characteristics, as with other 'big hitters', is the sheer length of his field experience. Such people may collect little in terms of number of specimens, but decades of experience enable them to be selective in what they collect.

These collectors adopted similar methods for floristic and taxonomic research and for the creation of seed banks. They focused on individual species in fruit or flower within habitats. In the 1990s William Hawthorne adopted a different approach to field collection in his investigations of West African plant diversity. In collaboration with African colleagues, he collected everything in an area, whether it was fertile or sterile. Where an assiduous collector using traditional techniques might collect two dozen specimens a day, Hawthorne and his colleagues collected hundreds. This approach presents enormous logistical challenges for plant drying, transport and handling but can provide a complete snapshot of plant diversity in a particular habitat at a particular time.

Collectors who augmented the university's collections before the 1980s often benefited from being able to collect over long periods, roaming across continents at will. The wide-ranging, months-long collection expeditions led by collectors such as Frank White, which traversed Africa between the 1950s and the 1970s, are inconceivable today. Politics, attitudes and priorities have changed. Collectors of the past generally took what they wanted with little concern for anything other than ensuring that they had obtained the best material available and that it was returned to Britain in the best possible condition. Today, permits and permissions are of equal concern to those who augment to the university's collections.

The collectors who have contributed to Oxford's collections have ranged the planet for plants. The concentration of the products of their efforts in Oxford – whether or not they were desired by the university – is as much an accident of history and philanthropy as of planning. The motivations of these men (and they have been mainly men) have been diverse but most would have wanted the material they collected to be used. Collections as trophies have no scientific value.

The potential of these collectors' work can be limited by a lack of collection details (e.g. Sibthorp's eastern Mediterranean specimens) or by the practices of collection owners after they receive the specimens (e.g. William Sherard and Henry Fielding). Rectifying such problems can be a major challenge that puts some researchers off using specimens.

Edward Lhwyd, FRS (*c*.1659–1709),[75] was praised as the best naturalist in early eighteenth-century Europe by the period's premier collector, Hans Sloane. As an illegitimate child, Lhwyd had his early botanical education overseen by his father's gardener Edward Morgan, who was associated with the Westminster Physic Garden.[76] By 1682, Lhwyd was in Oxford, where he became part of the fledgeling Oxford Philosophical Society, a scientific group led by Robert Plot. This intellectual network, which included men such as the English naturalists John Ray and Martin Lister, was to shape the rest of Lhywd's life.

In 1687 Plot made Lhwyd his assistant at the Ashmolean Museum. Despite his fear that Elias Ashmole would overlook him, Lhwyd eventually succeeded Plot.[77] With his promotion, Lhwyd's circle of correspondents increased as he catalogued British fossils, which he believed to be natural products of the earth rather than evidence of extinct life forms.

Like his mentor before him, Lhwyd conceived of a vast work, in his case a natural history of Wales. The research took him on extensive collecting expeditions through Cornwall, Wales, Ireland, Scotland and even Brittany. Lhwyd was also a pioneering linguist of the Welsh language. However, when he published the first volume of his project in 1707, his audience was disappointed; they had expected more from his field research.

Lhwyd died penniless in the basement of the Ashmolean Museum (now the History of Science Museum), and is commemorated in the genus *Lloydia*, the so-called Snowdon lily.[78]

Mark Catesby, FRS (1683–1749),[79] was one of the 'Procurers of Plants' for Philip Miller, the head gardener at Chelsea Physic Garden. He had no connections to Oxford, but most of the plant specimens he collected during his pioneering exploration of the southern part of colonial North America between 1722 and 1726 became part of the university's collections.[80]

Catesby was born in Essex. He lived with relatives in Virginia during the 1710s, and proved his worth as a collector of American plants to English natural historians, and was also said to '[design] and [paint] in water-colours to perfection'.[81] When he returned to England, his skills came to the attention of the influential naturalist and botanical patron William Sherard, who organized a group of high-profile British natural historians and horticulturalists to support Catesby's explorations of the southern American colonies in the 1720s.

Catesby returned to England in 1726, having sent thousands of herbarium specimens and living plants, sometimes packed in precise and ingenious ways, to his English sponsors. In London he embarked on the protracted process of preparing the illustrations and text for *The Natural History of Carolina, Florida and the Bahama Islands* (1729–47). He also became a strong advocate for the cultivation of North American trees in England.

Catesby is commemorated in the name *Catesbaea*, a small genus of shrubs from the West Indies and Florida.

John Sibthorp, FRS (1758–1796),[82] explored the eastern Mediterranean, introduced the artistic talents of the Bauer brothers, Ferdinand and Franz, to British botanists and founded the Sibthorpian Chair of Rural Economy.

Sibthorp was born in Oxford to a wealthy, landowning, academic family. His father, Humphrey, was the second Sherardian Professor of Botany in the university. In 1778, while studying medicine at Edinburgh University, Sibthorp was introduced to Linnaean botany by John Hope. A decade later, he was giving similar lectures to his students at Oxford.

In the early 1780s Sibthorp acquired the resources to continue his botanical studies in Paris and Montpellier, before returning to Oxford in 1783 where his father resigned the Sherardian chair in his favour. Sibthorp returned to the Continent almost immediately. In Vienna he met Ferdinand Bauer, one of the finest botanical artists who has ever lived.

During much of 1786 and 1787, Sibthorp and Bauer explored the botany of the eastern Mediterranean, determined to give modern Linnaean names to the medicinal plants recorded by Dioscorides in *De materia medica*. As they explored, they collected plants and Bauer made sketches. On his return, Sibthorp settled into academic life at Oxford. He tried to modernize the Botanic Garden, and to recover botany's reputation in the university after decades of neglect by his father. In March 1794 he travelled to Greece, but was forced to return to England fourteen months later. He died in Bath.

Sibthorp did not live to see the culmination of his work, but he left money in his will for its publication. Others such as James Edward Smith edited Bauer and Sibthorp's work and saw it through to publication as the unillustrated *Florae Graecae prodromus* (1806–16) and the fabulously illustrated *Flora Graeca* (1806–40).

4 BUD
Naming and Classifying

Research gives meaning to a collection. The acquisition of objects merely to swell the size, and perhaps status, of a collection is an exercise in vanity. Botanical research in the university over the past four centuries cannot be divided into neat categories, as research philosophies and methodologies have transformed the questions that can be asked of collections in all their diversity of forms. The first three centuries of research were largely concerned with questions of naming, describing and arranging the world's plants.

Accurate names are central to the uses we make of plants. Distinguishing between plants may make the difference between life and death. Furthermore, being able to communicate about the specific properties of plants depends on giving them unique names. For peoples as diverse as the Egyptians and Assyrians, through the Chinese and Indians, to the Greeks, names were crucial. Indeed, one of Adam's first challenges in Eden, according to the biblical book of Genesis, was taxonomic: to name 'cattle, ... fowl of the air and every beast of the field'.[1] Moreover, Christian doctrine viewed all organisms on the planet as God-given and immutable.[2]

Names are essential if information about plants is to be communicated between people and across generations, within a culture or between cultures. Importantly, names should be applied unambiguously, with one name referring to one thing. In early modern natural history, a system of naming soon developed based on short Latin descriptions (polynomials), for example *Jacobaea Sicula Chrysanthemi facie* ('Mayweed-like Sicilian ragwort'). Today this plant is known scientifically as *Senecio squalidus* (Oxford ragwort), using the binomial naming system formalized by the Swedish botanist Carl Linnaeus in 1753.

By the early seventeenth century, exploration of the globe had challenged the idea that all plants created by God were known to

Watercolour of *Daucus guttatus* by Ferdinand Bauer (*detail; see p.117*)

THE CARROT

Carrot (*Daucus carota*) is a very variable, widespread species whose natural distribution stretches from the Atlantic coast of Great Britain and Ireland, through Europe and the Mediterranean, to Central Asia. The species is readily recognizable by its highly divided, distinctively scented, fern-like leaves and its clusters of tiny, white flowers, arranged in parasols (umbels), that arise from a nest of finely divided bracts. Underground, wild carrot has a small, tough, highly branched, white tap-root. Cultivated carrots, in contrast, have swollen, unbranched tap-roots in a rainbow of colours – purple, yellow, red, orange and black.

During our evolutionary history, we soon learned to use plants' defence chemicals as medicine, and their starch, sugar, protein, vitamin and mineral contents as food. Carrots are valued and cultivated for the sweetness of their swollen roots. Carrot relatives, such as cumin (*Cuminum cyminum*) and parsley (*Petroselinum crispum*), are valued as herbs and spices, while hemlock (*Conium maculatum*) and cowbane (*Cicuta virosa*) are highly toxic.

Consequently, the ability to distinguish between these species could mean the difference between life and death.

Carrots are biennials: they take two years to mature. In the first year, cells in the tap-root swell with stored sugars. In the second year, these sugars fuel the formation of flowers and fruits. Cultivated carrots rarely flower, however, and are harvested at the end of the first year, when the tap-root sugar content is at its highest. Another relative of the carrot, whose tap-root is used as food is the parsnip (*Pastinaca sativa*). Classical and early modern authors did not distinguish clearly between carrots and parsnips.

Modern carrots were selected from wild carrots in Central Asia, where they 'practically invited themselves to be cultivated',[3] before spreading throughout Europe. During the seventeenth century, the kaleidoscope of carrot colours in western Europe was gradually reduced to one: European carrots became stereotypically orange. Subsequently, orange carrots followed Western empires as they colonized much of the globe.

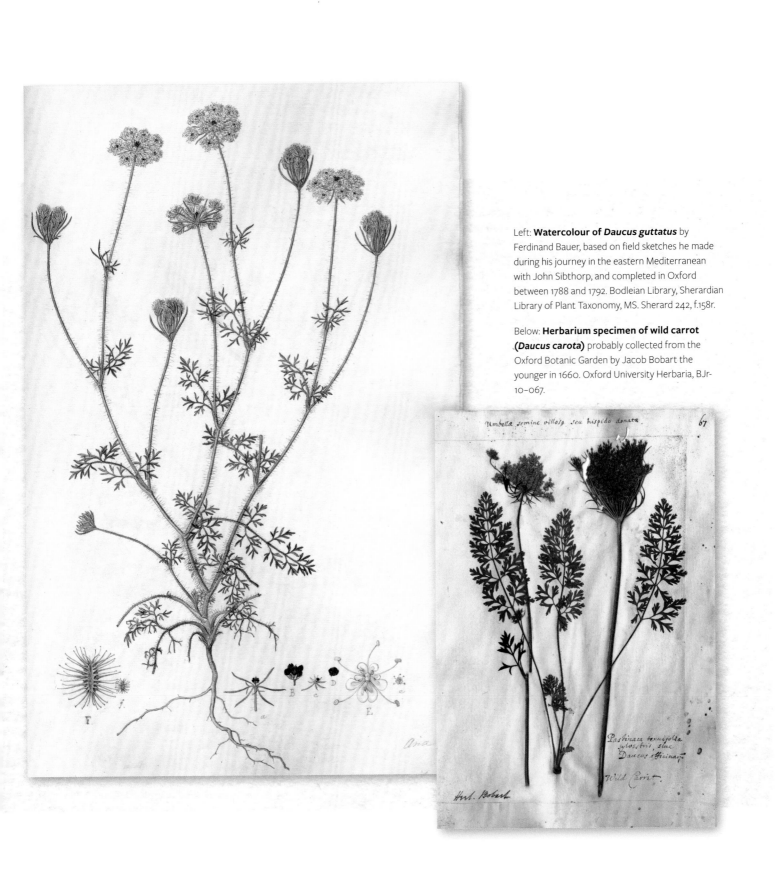

Left: **Watercolour of *Daucus guttatus*** by Ferdinand Bauer, based on field sketches he made during his journey in the eastern Mediterranean with John Sibthorp, and completed in Oxford between 1788 and 1792. Bodleian Library, Sherardian Library of Plant Taxonomy, MS. Sherard 242, f.158r.

Below: **Herbarium specimen of wild carrot (*Daucus carota*)** probably collected from the Oxford Botanic Garden by Jacob Bobart the younger in 1660. Oxford University Herbaria, BJr-10–067.

and mixed with the Oyl, and dropped into the Ears, easeth pains in them. The Root mixed with Bean-flower, and applyed to the Throat or Jawes that are inflamed, helpeth them, and the Roots or Berries beaten with hot Oxe-Dung, and applyed, easeth the pains of the Gout. *Tragus* reporteth, that a dram or more, if need be, of the spotted *VVake-Robin*, either green or dryed, being beaten, and taken, is a most present and sure Remedy for Poyson, and the Plague. The Juyce of the Herb taken to the quantity of a spoonful, hath the same effect; to which if there be a little Vineger added, as also to the Root aforesaid, it somewhat allayeth the sharp biting tast thereof upon the Tongue. The green Leaves bruised, and layd upon any Boyl or Plague-sore, doth wonderfully help to draw forth the poyson. A dram of the Powder of the dryed Root, taken with twice so much Sugar, in the form of a licking Electuary, or the green Root, doth wonderfully help those that are pursie and short winded, as also those that have the Cough; it breaketh, digesteth, and riddeth away Flegm from the Stomack, Chest, and Lungs. The milk wherein the Root hath been boyled, is effectuall also for the same purpose. The said Powder taken in Wine, or other drink, or the Juyce of the Berries, or the Powder of them, or the Wine wherein they have been boyled, provoketh Urine, and bringeth down Womens Courses, and purgeth them effectually after Child-bearing, to bring away the after-birth, and being taken with Sheeps milk, it healeth the inward Ulcers of the Bowels. The Leaves and Roots also boyled in Wine with a little Oyl, and applyed to the Piles, or falling down of the Fundament, easeth them ; and so doth the sitting over the hot fumes thereof. The fresh Roots bruised, and distilled with a little milk, yieldeth a most soveraign water to cleanse the skin from skurf, freckles, spots, or blemishes whatsoever therein. The fresh Roots cut small, and mixed with a Sallet, will make excellent sport, with a sawcy sharking guest, and drive him from his over-much boldness, and so will the Powder of the dry Root, strewed upon any dainty bit, that is given him to eat : For either way, within a while after the taking it, it will so burn, and prick his mouth and throat, that he shall not be able to eat any more, or scarce to speak for pain : The green leaf biteth the Tongue also. To take away the stinging of either, give the party so served new milk, or fresh butter. This Plant should be Venereous by its Signature.

Sisyrinchium. Spanish Nut.

Bulbosa Iris.

CHAP. XXXIII.

Of the Flower de Luce.

The Names.

THe Greeks call it, 'Ιελς as alfo 'Ιτελς, *quafi* Sacra, whereupon fome have tranflated it *Confecratrix*, all great and huge things being counted by the Ancients to be Holy ; but it was called *Iris, à cæleftis Arcûs fimilitudine, quam flores ejus reprefentant* ; from the Rainbow whofe various colours the flower thereof doth imitate. There have been fome heretofore that made a difference between *Iris* and *Ireos*, according to the Latine verfe extant thereof, which is this, *Iris purpureum florem gerit, Ireos album* ; but this is an errour proceeding as fome fuppofe from the Greek word Λειειον which fignifies a white Lilly, and by cafting away the firft letter becomes ειειον *er* ειειϑ by changing the laft fyllable, as if the Lilly and the *Iris* were all one, of which moft Authors make a diftinction : It is called *Radix Marica*, becaufe it is excellent for the Piles ; and fome have called it, *Radix Naronica* of the River *Naron*, by which great ftore doth grow. The knobbed *Iris* is called of *Matthiolus, Hermodactylus Verus*, becaufe the roots are like unto fingers;and from him divers did fo call it,but moft erroneoufly,it being a wild kind of flower de luce,as *Dodonæus* truly affirmeth. *Gladwin* which is a kind hereof, alfo is called in Greek, Ξϑεις *Xyris ob Folii fimilitudinem, quafi Raforium cultrum, aut novaculum dixeris* ; becaufe of it Swordlike or fharpedged Leaf, and in Latine *Spatula*,or *Spathula fætida* ; for *Spatha*, is taken for a fword as *Gladium* is ; and I have heard it called Roft Beef, for that the leaves being bruifed fmell fomewhat like it. The Flowerdeluce is called in Englifh *Iris* but moft commonly *Orris*.

The kinds.

So many of the forts as I find fet down in *Parkinfons* Theater of Plants, I here fet down ; which are eight. 1. The greater Broad leafed Flowerdeluce, 2. The greater Narrow leafed Flowerdeluce. 3. Portingall Flowerdeluce. 4. Broad leafed dwarf Fowerdeluce. 5. Stinking Gladwine. 6. The firft broad leafed bulbed Flowerdeluce of Clufius. 7. The greater bulbed Flowerdeluce. 8. The leffer bulbed Flowerdeluce ; to which I adde. 1. *Iris tuberofa* the knobbed Flowerdeluce;2. The common Flowerdelucer;3. Water flags or wild Flowerdeluce.

The Form.

The Common Flowerdeluce hath long and large flaggy leaves, like the blade of a fword with two edges, amongft which fpring up fmooth and plain ftalks, half a yard long,or longer, bearing flowers towards the top, compact of fix leaves joyned together : whereof three that ftand upright are bent inward one toward another, and in thofe leaves that hang downwards there are certain rough and hairy Welts, growing or rifing from the nether part of the leaf upward, almoft of a yellow colour, The Roots be long, thick and knobby, with many hairy threds hanged thereat ; but being dry it is without them, and white.

The Places and Time.

Thefe Fowerdeluces aforementioned, and many more, though they grow **naturally** in *Africa*, *Greece*, *Italy* and *France*, and fome in *Germany* : yet they

K 2   are

William Cole was a Fellow of New College, whose fantastical *Adam in Eden, or, Natures Paradise* (1657) elaborated upon the doctrine of signatures, which asserted that uses could be divined from forms; thus knobbly iris roots were effective treatment for tumours. The copy shown is augmented with marginal sketches and hand-coloured woodcuts clipped from Rembert Dodoens's *A Nievve Herball* (1578). Bodleian Library, Sherardian Library of Plant Taxonomy, Sherard 424, 66–67.

humans. Each voyage of exploration or trade that returned to Europe brought seeds, specimens and reports of plants that were not mentioned in either biblical or classical texts. Even if these reports named all the plants and supplied descriptions of them, alphabetical lists of names provided limited botanical knowledge.

Plants were commonly arranged into classifications based on habit, habitat or utility, but these were hardly of much general use. Experimentation with methods of arrangement saw people publishing lists of plants found in specific gardens, personal herbaria or parts of the world, or descriptions arranged according to species use, for example, in horticultural manuals or in herbals. Such volumes lacked the universality of a common arrangement. In the seventeenth and eighteenth centuries the focus shifted to general classification systems based on the reproductive parts of the plant – the flowers and the fruits. These would crystallize in the sexual classification system of Carl Linnaeus, an initially controversial system that was viewed with scepticism by the first Sherardian professor, Johann Dillenius.[4]

Broadly defined, taxonomic research has been a more or less constant part of plant sciences in the university since the seventeenth century. In the latter half of the century, the focus was on the naming and classification of flowering plants. In the first half of the eighteenth century, this shifted to the classification of fungi, mosses and ferns, and the creation of a global list of plant names. The dormancy that had characterized botanical research in the university in the latter half of the eighteenth century came to an end with Charles Daubeny's appointment in the 1830s, although taxonomic research did not regain its former standing until the 1950s.

Robert Morison's *New Universal Herbal*

By 1670 the University of Oxford was well placed to contribute significantly to the development of a universal plant classification system. A botanic garden in the care of Jacob Bobart the elder was well established, the Bobarts were drying plants for incorporation into their herbarium and Robert Morison had been appointed Regius Professor of Botany – the first such position in any English university. During exile in France, where he was employed by Gaston, Duke of Orléans, Morison had thought about

Late seventeenth-century oil portrait of Robert Morison, the first professor of botany in Britain, probably by William Sunman. University of Oxford, Department of Plant Sciences.

the best General Method taken from Nature itself, of digesting all Plants, and reducing them to certain Classes or Heads according to the difference of their Seeds, Podds and Flowers; by the advantage whereof the Study and Remembering of Plants may be much facilitated, and the Contemplation thereof among all sorts of Men exceedingly promoted.[5]

Once he had slaughtered some sacred botanical cows from earlier decades – including Jacques Daléchamps's habitat-based classification, John Parkinson's use-based classification and Basil Besler's seasonal classification – Morison summarized his approach in the *Praeludia botanica* (1669). Morison even made detailed criticisms (which he called 'hallucinations') of the Swiss botanical brothers Gaspard and Jean Bauhin, which coloured reactions to his work for the rest of his life. Morison held that plant classification must be based on characteristics that unite groups of plants; for example, buttercups (*Ranunculus*) all have flowers with five parts and pointed, compressed fruits, regardless of their habitat, properties or leaf shape.[6]

Morison's classification, called the *Sciagraphia* ('first draft'), was based on 'fruits' (no clear botanical distinction was made between fruits and seeds in the seventeenth century), hence Carl Linnaeus' classification of Morison as a 'fructist'. The system came from his study of the 'Book of Nature', while its focus on fruits had divine authority: 'And God said, Let the earth bring forth grass, the herb yielding seed, and the fruit tree yielding fruit after his kind, whose seed is in itself, upon the earth.'[7] The system, never published in its entirety during Morison's lifetime, first divided plants according to whether they were trees, shrubs, sub-shrubs or herbs, in the manner of Theophrastus, and then by the form of their fruits.[8] In Oxford, Morison had the opportunity to realize his ambition of producing a universal classification system, which had been thwarted by the death of his patron, the Duke of Orléans, in 1660. Work could now begin in earnest on the *Plantarum historiae universalis Oxoniensis*.

Despite being the king's botanist, he was not well remunerated, so Morison looked elsewhere for funding for his project.[9] To garner enthusiasm, in 1672 he published *Plantarum umbelliferarum distributio nova*, a folio volume about members of the carrot family. In addition to detailed plant descriptions, there were diagrams showing the

Copperplate engraving, sponsored by Peter Mews, vice chancellor of Oxford University, showing a comparison of fruits in Robert Morison's *Plantarum umbelliferarum distributio nova* (1672), the first published taxonomic monograph. Bodleian Library, Sherardian Library of Plant Taxonomy, Sherard 726, t.1.

Tab. I. Icon.

TABULA GENERALIS ICONUM SEMINUM UMBELLARUM.

Auspicijs R. viri D.ni Petri
Mews. LL.D. Coll-D. Io.B.Præs:
Dec:Roff:& Univ.Oxon.Vicecan:

similarities of plants within the groups he defined and twelve pages of copper-plate illustrations by unnamed artists. The engraving of each plate was sponsored by senior members of the university and colleges such as John Fell, the vice chancellor at the time of Morison's appointment and future bishop of Oxford, and Robert South, the university orator who had expressed his scepticism of the Royal Society when the Sheldonian was opened.[10] This volume revealed the extent and quality of his ambition. He wanted his *Historia* to appear in three parts: part 1 was to include woody plants, while parts 2 and 3 would focus on herbs. He started with part 2, in his view the most complex part of the scheme, because he wanted to ensure that if he died before the project finished it would not be completed by an 'incompetent person'.[11]

In 1675, in a 'Proposal to Noblemen, Gentlemen and others', Morison asked subscribers to give him five pounds each to support the preparation of each plate of the *Historia*. In return, they would get 'an Honourable memorial … by engraving their Coat of Arms on their respective Plates (as is already done to others in the foremention'd Essay [*Plantarum umbelliferarum*], as likewise in an hundred Plates and upward, of the five Sections now in hand)', a copy of the final printed volume and, when they paid their subscription, 'one of his Specimens Umbellarum'.[12] He had 'good hopes, he shall by the Assistance and Encouragement of the Generous, be enabled to give good satisfaction to the Curious therein'. The costs of engraving the plates, which have survived to the present day and are preserved by the Bodleian Libraries, proved far greater than the funds raised.[13]

Morison lived to see the completion of only part 2 of the *Historia*, published in 1680. Part 3 was completed by Jacob Bobart the younger, a protégé of Morison and one of the few men to adopt his classification system, albeit in a modified form, and published in 1699. Part 1 was never published. The *Historia* as an illustrated catalogue of plant diversity was never realized.

Morison's classification system was soon eclipsed by that of his contemporary, the Essex-based naturalist John Ray. Ray's *Methodus plantarum nova* (1679) was published the year before Morison died, and would eventually form the basis for his three-volume magnum opus, *Historia plantarum* (1686–1704). Many reasons have been advanced for the rejection of Morison's classification system by naturalists after its publication. These include his death before the whole system was published, his poor relationship with John Ray and his vanity in refusing to acknowledge his debt to earlier botanists, particularly the sixteenth-century Italian botanist Andrea Cesalpino.[14]

Although Morison's classification system was all but ignored, botanists such as William Sherard, Richard Richardson and Robert Sibbald, co-founder of the botanic garden in Edinburgh,[15] made use of some of the names Morison gave to plants. Most importantly for botany, in the latter half of the eighteenth century, Linnaeus used Morison's names in his *Species plantarum*, although in Linnaeus' view Morison 'followed the thread of nature, ties his own thread of Ariadne into Gordian knots, which can be untied only with the sword'.[16] In a long undated letter to the Swiss botanist Albrecht von Haller, Linnaeus went further:

> Morison was vain and puffed up with conceit, ... yet he cannot be sufficiently praised for having revived system, which was half expiring. If you look over Tournefort's genera, you will readily admit how much he owes to Morison ... All that is good in Morison is taken from Caesalpinus, from whose guidance he wanders in pursuit of natural affinities rather than of characters.[17]

Morison's most lasting scientific contribution is not the *Historia*, however, but the *Plantarum umbelliferarum*, where he presents a detailed analysis of a single, taxonomically defined group of plants – the carrot family. He set the stage for the taxonomic botanical monograph, where all knowledge about a discrete group of plants is collected, critically reviewed and synthesized.

Monographs

Following the publication of Morison's *Plantarum umbelliferarum*, the next monograph to be completed by an Oxford botanist, albeit two years before he arrived at Oxford as the Sherardian Professor of Botany, was by Johann Dillenius. In the *Hortus Elthamensis* (1732), Dillenius published an account of sixty-three known members of a group of African succulent plants called 'Mesembryanthemum'; today, they are divided across various genera in the family Aizoaceae.

The *Hortus* itself describes exotic plants cultivated by the wealthy, early eighteenth-century apothecary James Sherard at Eltham, Kent, and was started by Dillenius in 1724. Over eight years, Dillenius drew and engraved 325 folio-sized copper-plates and described 418 plants. He resented the time he spent on the project, and having to make it look 'bigger and more pompous' at Sherard's behest.[18] In turn, Sherard complained that Dillenius' chief concern was 'to improve and advance the knowledge of botany' rather than to promote the wonders of his garden.[19] Despite pressure from his employer, Dillenius succeeded in producing a classic work of eighteenth-century horticulture,[20] which remains a modern botanical resource. High-quality taxonomic monographs remain scientifically relevant long after their authors are dead.

Carl Linnaeus visited Oxford to see Dillenius in 1736. Following a famously frosty initial meeting, the two men remained regular correspondents until the end of Dillenius' life, and Dillenius sent Linnaeus a copy of the *Hortus*; nearly 90 per cent of the *Hortus* plates are cited in Linnaeus' *Species plantarum*.[21] In fact, over a hundred of the plates are taxonomic types. Types are the objects that people use when new species are described for the first time; examination of type specimens is essential before new species are named. Furthermore, the models for many of Dillenius' illustrations are specimens preserved in the Oxford University Herbaria.

In modern times, monographic botanical approaches were established at Oxford by Frank White in the 1950s, especially in the ebony family (Ebenaceae).[22] Monography was taken up by David Mabberley in the 1970s, working on genera in the mahogany family (Meliaceae), and most comprehensively by Robert Scotland in the twenty-first century, on sweet potatoes (*Ipomoea*).[23] In contrast to the solo monography that characterized the work of Morison and Dillenius, subsequent monographers at Oxford have worked, to a greater or lesser extent, collaboratively with teams of postdoctoral researchers and postgraduate students. In addition, these research groups cooperate with researchers, collection curators and fieldworkers worldwide. Oxford – and individual academics – became part of wider local and global taxonomic research communities.

Johann Dillenius' 'lower plants'

Before Johann Dillenius, often called 'the father of British bryology', became the first Sherardian professor, his botanical reputation was based

The bat-pollinated *Calliandra houstoniana* var. *anomala* (large-flowering sensitive plant) from Mexico and northern Central America is incorrectly associated with hummingbirds and Jamaica in Robert Thornton's *Temple of Flora* (1807). This aquatint was made by Joseph Stadler, based on an original oil painting by Philip Reinagle, in 1799. Bodleian Library, Sherardian Library of Plant Taxonomy, 582 LI 13.

on three taxonomic works. The *Catalogus plantarum sponte circa Gissam nascentium* (1718), an account of more than 1,300 plants (including mosses and fungi) growing around the Hessian town of Giessen, showed that he could produce Floras, that is, books about the plants growing in geographically defined areas. His editing of the third edition of John Ray's *Synopsis methodica stirpium Britannicarum* gave him the credentials of someone who understood classification systems. His work on the *Hortus Elthamensis* (1732) was an example of his monographic work.

In Oxford, Dillenius returned to his interest in 'lower plants', that is, mosses, lichens and algae, which he 'discovered and sorted out with stupendous industry'.[24] As a young man, he had tried to explain how mosses reproduce,[25] and at Oxford he produced his most original scientific work, the *Historia muscorum*, although it sold poorly during his lifetime.[26] In 576 pages of Latin text and eighty-five plates drawn and engraved by himself, Dillenius attempted something that had not been done before; he brought order to the confusion of names surrounding the 'lower plants'. The work done by Bobart the younger for Morison's *Historia* (1699) had only created confusion because of the difficulty of reliably separating species using gross morphological characteristics.

Dillenius collected many British specimens and collaborators also sent him samples from Germany, Holland, Poland, Russia, Sweden, Greenland, the Carolinas, Pennsylvania, Virginia, Jamaica and the Bahamas, and even Patagonia. He also included all the herbarium specimens at his disposal, and did not scruple to remove specimens of interest from the university's collections and add them to his personal herbarium.[27] Using the facilities available to him, Dillenius examined the specimens meticulously, trying to reconcile what he saw with what was reported in the literature. Over 600 species are described in the *Historia muscorum*. Dillenius' foresight in associating these descriptions with herbarium specimens in his personal collection contributed to the lasting importance of the *Historia muscorum*, especially for the study of lichens worldwide.

Linnaeus did not appear to examine 'lower plants' critically for his *Species plantarum*, but rather made wholesale use of Dillenius' *Historia*. Consequently, Linnaeus' work has been criticized because of the 'vague & uncertain'[28] characters chosen by Dillenius to separate groups (genera). In the case of mosses, it was the work of the German botanist Johann Hedwig in the late eighteenth century that formed the basis of

Herbarium specimens of the liverwort *Marchantia* used by Johann Dillenius when he wrote *Historia muscorum* (1741). Today, the genus is used as a model plant to study developmental biology. Oxford University Herbaria, HM_166.

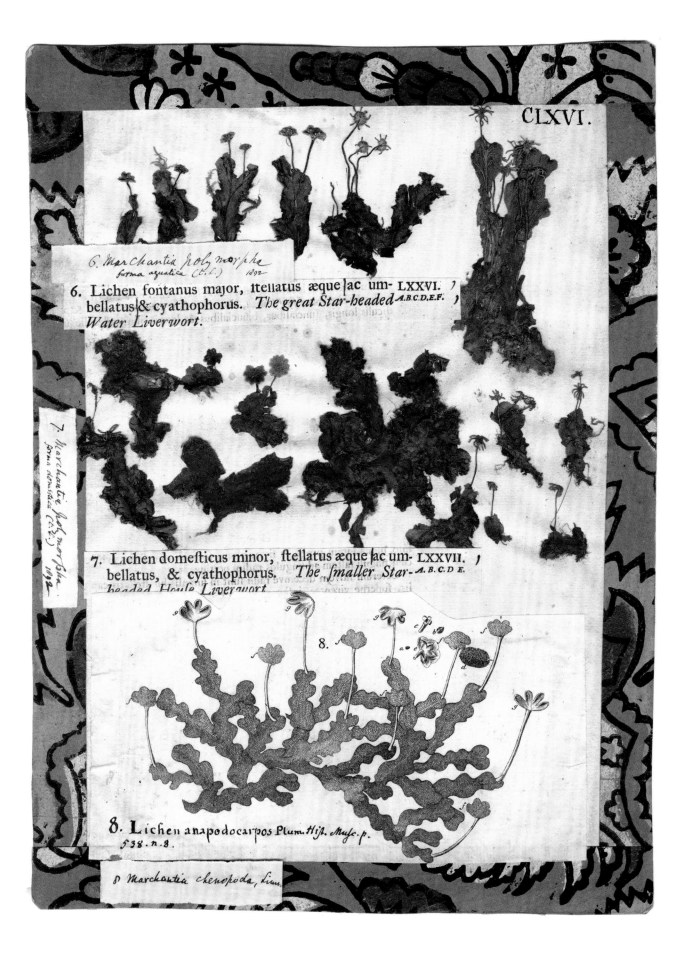

CLXVI.

6. *Marchantia polymorpha*
forma aquatica (E.L.) 1892

6. Lichen fontanus major, ſtellatus æque ac um- LXXVI. }
bellatus & cyathophorus. *The great Star-headed* A.B.C.D.E.F. }
Water Liverwort.

7. *Marchantia polymorpha*
forma domestica (E.L.) 1892

7. Lichen domeſticus minor, ſtellatus æque ac um- LXXVII. }
bellatus, & cyathophorus. *The ſmaller Star-* A.B.C.D.E. }
headed Houſe Liverwort.

8.

8. Lichen anapodocarpos Plum. Hiſt. Muſc. p.
538. n. 8.

8 *Marchantia chenopoda, Linn.*

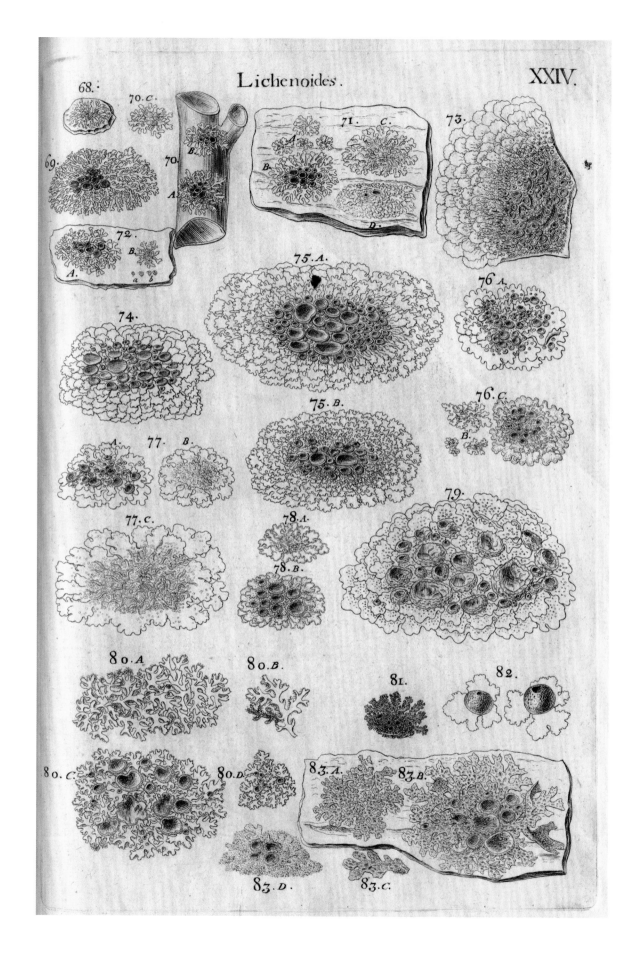

modern moss naming systems, although the characters he selected 'are so small that they require the Assistance of a Microscope of considerable Powers to examine them with precision & count them with Accuracy'.[29] Such facilities were not available to Dillenius or Linnaeus.

Linnaeus held the taxonomic work of Dillenius in high regard. He dedicated his *Critica botanica* (1737) to Dillenius and stated that 'there is nobody in England who understands or thinks about genera except Dillenius'.[30]

William Sherard's list of plant names

By the end of the sixteenth century, natural philosophers had published tens of thousands of disconnected names for the plants they studied. Confusion and contradiction reigned as the Swiss botanist Caspar Bauhin began the laborious, four-decade-long task of distilling these names to approximately 6,000 species and their synonyms in a pre-Linnaean version of the current global World Flora Online.[31] The publication of Bauhin's *Pinax theatri botanici* (1623) was crucial to communicating accurately and scientifically about plants in the seventeenth century. It provided a common basis for answering the question of how many plant species there were on earth. By the end of the seventeenth century, however, Bauhin's *Pinax* was inadequate, and a revision was needed.

Joseph Pitton de Tournefort in Paris thought that William Sherard had the intellectual and personal qualities to undertake such a revision.[32] Sherard's *Pinax* is formed around a cut-up copy of Bauhin's *Pinax*, to which he added new synonyms and references, and intercalated new species names. Notes on the sources of unpublished names compete for space with the names of plants collected from or seen in gardens across Europe. Thousands of entries were added to Bauhin's arrangement as Sherard and his collaborators searched European botanical collections for new plant names and tried to determine whether they were original or merely synonyms of existing ones.

The project occupied the last decades of Sherard's life, and the later careers of Dillenius and his successor, Humphrey Sibthorp. The project died with Sibthorp in 1797 and has remained incomplete and unpublished; the 6,215 densely packed, handwritten pages of the Sherardian *Pinax* are a testament to the Sisyphean task Sherard had begun.[33] With hindsight, the devotion of Oxford-based botanists to

Copperplate engraving of lichens by Johann Dillenius, based on his own illustrations for his *Historia muscorum* (1741). Bodleian Library, Sherardian Library of Plant Taxonomy, Sherard 442, t.XXIV.

Gnaphalium montanum, purpureum Ac. R. Par. ico.

* Elichrysum montanum flore rotundiore, sub purpureo, suaverubente candido. Item Elichrysum montanum longiore folio & flore, purpureo & albo tuj. art. Cat. Gif. 160. non enim specie, sed sexu tantum different.
Hispidula Schwed. Lib. 4. Class. 1. Gnaph. mont. flore suaverubente M
Gnaphalium flore albo & rubro Reh. Add. 35. Cult. 437.
Gnaphalium montanum flore rubro & purpurascente Merr. Pin.
Lagopus 2. Hasenpfoetlein Trag. L. 1. C. 88. p. 266. & C. 109. p. 331. 332. (fij. bon.)
flore candido, purpureo & roseo. Gnaph. sylvestre flore purpureo Vesl. H.
Auricula muris 4. seu Hasenpfoetlein id. L. 1. C. 93. p. 279.
Gnaphalium minus album Schwenck. Cat. 88.
Gnaphalium montanum flore rotundiore albo Kyll. Vir. san. 57.
Elichrysum montanum flore rotundiore candido Just. r. h. 45.
Boerh. Ind. alt. 120. Vaill. comm. Ac. R. Sc. Ann. 1719. p. 293. n. 33.
Gnaphalium montanum flore rotundiora candido C. B. Pin. 126.
Gnaphalium Alpinum flore albo Vesl. Hort. Cat.
Gnaphalium montanum album Lob. icon. 482. Tab. Hist. lib.
109. icon. 391. Gnaph. montan. flore albo Munt. Cult. 437.
Gnaphalium montanum flore rotundiore suaverubente Kyll. Vir. san. 57.
Elichrysum montanum flore rotundiore, suaverubente Just. r. h.
Vaill. comm. Ac. R. Sc. Ann. 1719. p. 294. variet. spec. 33.
Gnaphalium montanum flore rotundiore suaverubens Lob.
Gn. mont. folio & flore longiore Tab. 333. & ejus praejus leorsum
icon. 483. Gn. mont. flore rotundiore Just. ord. 3. tab. 333. Ci Tab. Hist. lib. 2. 10.
icon. 392. Gnaphalium minus suaverubens Schwenck. Cat. 88.
Elichrysum montanum, flore rotundiora variegata Just. r. h.
Vaill. comm. Ac. R. Sc. Ann. 1719. p. 294. var.
Boerh. Ind. alt. 120. suaverubens.
Gnaphalium montanum, variegatum Eyst. ord. 3. pl. vern. 16
2. & 3. Gnaph. mont. folio rotundiore variegatum Vir. Act. t. 933. (ex ...)
Gnaphalium montanum, fl. variegato, nostrum Chlor. Goth.
Fl. Wehsburg.

= Hist. pl. Ag. saxt. ico.

* Gnaphalium montanum fl. candido, longiore Ac. R. Par. ico.
————— nudum Suer. Cat.

Gnaphalium montanum alatum, fl. oblongis, amoena purpureis, morianum pl. mant. g.

Elichrysum montanum, longiore et folio et flore maiore rubente suppl. Fl. Pruss. Vaill.
Comm. Ac. R. Sc. Ann. 1719. p. 294. venetis specie 34.

* Elichrysum montanum, longiore et folio et flore sulphureo suppl. Fl. Pruss.

Elichrysum montanum flore rotundiore subpurpureo Inst. r. h. 453.
Raii Comm. Ac. R. Sc. Ann. 1719. p. 294. Vaillant. spec. 33.

Gnaphalium mont. um sal subpurpureo florum Cat. H. Cat. a Turre

C. B. Pin. 263. Inst. r. h. 453. Tourn.
Inst. 2. c. b. 487.

I. Gnaphalium montanum flore rotundiore.
Lagopus secundus, Trag...
Auricula muris, Lon...
Pilosella minor, Fuch. Dod. gal. Lugd. Thal. 83
Lagopiron Hippocratis, Gef. animal.
Gnaphalij genus, Gef. col.
Gnaphalium montanum purpureum, Scalbum,
Ad Lob. Tab. Ger. 516.
Gnaphalium montanum suaverubens, &
Gnaphalium montanum variegatum, Eyst.
    Variat floribus candidis, subpurpureis, roseis, ex candido & rubro mixtis.

Gnaphalium montanum purpureum, Ger.
Umbr. 641. Pilosella minor Sw. Phyt. 86. Clus.
floribus candidis & purpureis Pempt. 18. fig. 2.
1098. Gall. raii. 2. 2.
... 96.

Hist. lib. 2. 109. icon. 392.
R. Hist. 1. 243.

Gnaphalium montanum um-
gatum Turr. H. Patav.

Gnaphalium montanum suaverubens Bod. a St. in H.

Gnaphalium montanum major fl. rot. C. B. f. Lob. icon. 468.

Pilosella major quibusdam, alijs Gnaphalij genus J. B. 3. 162.

Hispidula Rhodoei officinarum Jungermanni.

Elichrysum montanum, sive spas Cat. H. R. Monsp.

Idem floribus subpurpureis C. B. var. J. B. 3. 162. dasc.

Idem flore ex candido et rubro mixtis C. B. var. J. B. 3. 162. dasc.

Idem (majus) floribus luteis H. A. L. B.

Idem floribus sulphureis H. A. L. B.

Gnaphalium montanum sive spas Cat. Park. theat. Par. p. 375. des.

Chrysocome humilis montana folio rotundiore purpurea et alba Hist. Oxon. p. 3. 89. n. 32.
V. Com. Ac. R. Sc. Ann. 1719. p. 294. C. 453.

= Elichrysum montanum longiore et folio et flore purpureo dasc. r. h.
x Phytog. 511.

II. Gnaphalium montanum longiore & folio & flore. C. B. Pin. 263.
Gnaphalium montanum purpureum, & suaverubens, Lob. ico. Tab. Ger.
Gnaphalium montanum Dalechampij, Lugd.
Pilosella minor, Dod. Clus. hist. 334.
    Variat ut prior floris colore; hinc duæ icones apud Lob. Tab. Gerard.

Inst. r. h. 453. Tab. fi. 392.
Gnaphalium montanum purpureum Ger. 516. Um.
640. Lob. Ic. 1. 483. Belg. 568. (ic. melior.)
1116. Gall. Lob. 2. 18. Gnaph. minus purpureum Schw. Gnaph. Cat. 88. ubi it.
Pilosella minor quibusdam, alijs

Gnaphalij genus J. B. 3. 162.

Chrysocome humilis montana auctiore folio Hist. Oxon. p. 3. 89.

Elichrysum montanum longiore et folio et flore albo Inst. r. h. 453.

Gnaphalium montanum, longiore et folio et flore albo C. B. Pin. 263.

Gnaphalium montanum longiore flore carneo Kyll. Vir. dan. 58.

Gnaphalium montanum longiore flore suaverubente Kyll. Vir. dan. 58.

Gnaphalium flore sulphureo Hermanni R. Hist. 2. App. 1861.

Gnaphalium montanum, flore rotundiore, floribus sulphureis H. A. L. B.

Sherard's *Pinax* for half a century was a distraction from botanical developments in other parts of Europe. The problem of synonymy and the determination of the number of plant species remains an issue for modern botanists.[34]

The art of botanical illustration

Botanical illustrations – detailed, scientifically accurate representations of plants or plant parts – have been central to the presentation of botanical research in the university and institutions worldwide. The best botanical artists work in close collaboration with botanists, their mutual talents feeding off each other. In the eighteenth century, Johann Dillenius was the artist and engraver for his own publications. In the early twentieth century, the botany lecturer Arthur Harry Church brought his exceptional artistic talents to bear on illustrating his own publications and his undergraduate lectures.[35] In modern times, the most prolific and long-serving artist associated with botany at Oxford is Rosemary Wise. In a career spanning over fifty years, Rosemary has illustrated more than 14,000 plant species for the academic publications of generations of botanists, many of whom trained at Oxford before taking up posts nationally and internationally.[36]

The two best-known botanical artists associated with Oxford botany are Georg Dionysius Ehret and Ferdinand Bauer. Both went on to establish their reputations outside Oxford, each having experienced tense working relations with Humphrey and John Sibthorp, respectively. Ehret, briefly employed as superintendent of the Botanic Garden in 1750, has the dual distinctions of being the shortest-serving superintendent in the Garden's history and the only superintendent to be sacked.[37] He left no mark on Oxford as either gardener or artist.

In contrast, Bauer arrived at Oxford in late 1787, with John Sibthorp, to transform the sketches he had made during their exploration of the eastern Mediterranean into some of the world's finest botanical illustrations. Following John's death, these watercolours were eventually published as the *Flora Graeca*, one of the rarest and most expensive botanical books ever written.[38]

In Oxford, Bauer probably lived in the servant's quarters under the eaves of Sibthorp's home, Cowley House. Their mutual respect for each other's gifts did not mask the animosity between these two talented men. However, Bauer was a professional, contracted to Sibthorp to work

previous pages **Typical pages from William Sherard's *Pinax***, with printed clippings from Bauhin's *Pinax* surrounded by annotations in the hands of Sherard and Johann Dillenius. Bodleian Library, Sherardian Library of Plant Taxonomy, MS. Sherard 113, ff.14v–15r.

on botanical watercolours, so he had to adapt. His job was to turn the pages of densely packed pencil sketches, surrounded by their clouds of numbers representing colours, into watercolours. During the six years he was in Oxford, Bauer completed 966 life-sized watercolours of plants, 248 folio-sized watercolours of animals and seven folio-sized frontispieces for the *Flora*.

Bauer would have transferred his sketches to the drawing paper and then painted with solid colour according to the colour code, using a limited palette of pigments.[39] Working without an assistant, he would have prepared his own paper, ground his pigments and mixed his own paints, in addition to producing the final watercolours. The mechanical description fails to capture the sheer quality of Bauer's artistry and the care and precision with which he worked. By 11 June 1792, he had completed nearly 1,000 botanical watercolours, some based on sketches he had made five years earlier, of plants seen only once. On average he completed one watercolour every one and a half days.

During the protracted publication process of the *Flora Graeca*, the accuracy of Bauer's watercolours passed the detailed scrutiny of such eminent botanists as James Edward Smith, Robert Brown, the first keeper of the Botanical Department of the British Museum, and John Lindley, professor of botany at University College, London. Any minor errors perhaps reflect Sibthorp not supervising the work as closely as he should have done. In one case, Bauer drew cobwebs on a specimen in error for the plant's hairs. But Bauer's methods, together with his own innate abilities, enabled him to produce truly outstanding natural history watercolours within a strict scientific framework.[40]

Following the death of Sibthorp in 1796, and the stipulations in his will, work began on the publication of the *Flora Graeca*. Sibthorp was aware that enthusiasm for a project was not enough. The difficulties Morison and Dillenius faced in funding publication of their research at Oxford would have been familiar to him. Consequently, he left funds in his will to complete the publication, and an incentive to the university to see that it was done. Funds for a chair in rural economy would be released only once the *Flora Graeca* was published to the satisfaction of his executors.

The laborious task of synthesizing Sibthorp's field notes, written in a poor hand with poor ink on poor-quality paper, with unlabelled specimens, sketches and watercolours and draft notes, completed by

Sibthorp in Oxford, fell to James Edward Smith. The amount of synthesis and primary research Smith had to undertake for the manuscript to be suitable for publication led to quarrels over the authorship of both the *Flora* and the primary scientific work, the *Florae Graecae prodromus* (1806–16). All this work took place outside Oxford; on Sibthorp's death, the original materials left Oxford and did not return until 1840.

Notwithstanding the wrangling associated with its publication, the *Flora Graeca* named and described hundreds of new species from the eastern Mediterranean for the first time. However, by the time it was published, the science of plant classification had moved on and the Linnaean sexual system used in the book was redundant. Today, the *Flora Graeca* is known for the extravagance of its publication and the magnificence of the hand-coloured plates based on Bauer's watercolours: the botanist has been eclipsed by the artist.

Affordable Floras

The *Flora Graeca*, a sumptuous example of a Flora, is a book focused on the plants of a geographically defined area. During the eighteenth and nineteenth centuries, lavishly illustrated, large-format Floras were regularly published across Europe.[41] Such works were not directed at 'Cabbage-planters; but to the best refined'.[42] However, they attracted censure because such 'Expensive performances ... recommend themselves only to persons, who, with a taste for the polite arts, possess also the means of indulging it; and to public libraries, the archives of what is curious in a country'.[43] The irony is that Sibthorp complained of the inaccessibility of lavish botanical publications as he travelled through Italy on his eastern Mediterranean journey; the *Flora Graeca* proved to be one of these.[44] The wealth of such volumes in Oxford's botanical collections has been achieved through the generosity of private collectors such as John Sibthorp, Charles Daubeny, Sydney Vines and George Claridge Druce.

Druce himself was an enthusiast of a rather different type of Flora. Rather than being replete with lavish illustrations, these Floras are text rich, often densely written and focused on details of local plant distributions.[45] They often come with botanical keys to aid the identification of plants and, in the past fifty years, with detailed maps of species' distributions.

In Britain these works have their origins in Ray's *Catalogus plantarum circa Cantabrigiam nascentium* (1660), a catalogue of the plants growing

Hand-coloured copper engraving of *Iris germanica* based on field sketches and a watercolour by Ferdinand Bauer and published in Sibthorp and Smith's *Flora Graeca* (1808). Bodleian Library, Sherardian Library of Plant Taxonomy, Sherard 761, pl. 40.

Iris germanica.

around Cambridgeshire.[46] The first Flora of Oxfordshire, published in the manner of Ray's work, was the only book that John Sibthorp published during his lifetime, *Flora Oxoniensis* (1794). The master of Magdalen College School, Richard Walker, published the first Flora of Oxfordshire in English, *The Flora of Oxfordshire and its Contiguous Counties* (1833), although it provoked considerable criticism from subsequent researchers because of its inattention to detail.[47]

The two editions of Druce's *Flora of Oxfordshire* (1886, 1927) are model Victorian and Edwardian county Floras in the United Kingdom. Based on Druce's extensive personal knowledge of Oxfordshire's plants and of previous botanical exploration, they record minute distributions of species across the county. They are also idiosyncratic. By modern standards, Druce's recording in Oxfordshire is biased to species or areas that were of special interest to him. For most researchers, the effort involved in the publication of one county Flora is a lifetime's work; to produce two distinct editions is remarkable; but to write Floras for three other counties (Northamptonshire, Berkshire and Buckinghamshire) as Druce did is extraordinary. The most recent county Flora, *The Flora of Oxfordshire* (1998), was the product of decades of research and collaboration between the Natural Environment Research Council botanist John Killick, the Oxford-trained cryptogamist Roy Perry and Stan Woodell, a former lecturer in the Department of Botany and founder member of the Berkshire, Buckinghamshire and Oxfordshire Wildlife Trust.

The university's collections have contributed to our understanding of British plants and to the production of national Floras. However, direct involvement with Flora projects covering the whole of the United Kingdom has been limited. Dillenius' revision of Ray's *Synopsis methodica stirpium Britannicarum* became the standard account of British plants for the next fifty years. William Baxter's *British Phaenogamous Botany* (1834–43), with its detailed species accounts, had a promising start but was not completed. In 1953 the first edition of Clapham, Tutin and Warburg's *Flora of the British Isles* (subsequent editions 1962 and 1987) was published; it became the standard account of the British flora for the rest of the century. Arthur Roy Clapham and Edmund Frederic Warburg had direct association with botany at Oxford. From the 1930s to the mid-1940s, Clapham had a teaching position under the Sherardian professor Arthur Tansley, contributed to Tansley's seminal work *The British Islands and their Vegetation* (1939) and was the originator of the word 'ecosystem', which is often attributed to Tansley.[48] From the late 1940s, Warburg had a post in the Department of Botany as curator of the Fielding and Druce herbaria, where he excelled as a teacher and a field botanist.

During the twentieth century, the work of researchers in the department was not confined to the British flora. Through the meticulous cataloguing of national floras, researchers such as Joseph Burtt Davy and Frank White made fundamental advances in our understanding of the African flora. Similarly, contributions were made to the compilation of Floras in the Americas and Asia by other Oxford-based researchers. Unlike lavish volumes such as the *Flora Graeca*, these county and national Floras are not ornaments for library shelves; it is expected that they will get dirty and battered, as they are used for all manner of practical purposes, by all manner of people. This ideal of a botanical reference book that is accessible to all may be epitomized by David Mabberley's *The Plant-Book*, first published in 1987 and now in its fourth edition.

The contributions made by Oxford botanists to the global task of naming, describing and cataloguing the plants of the planet have been episodic. Morison's implicit championing of monographic botany in the mid-seventeenth century was a break with the herbal traditional in previous centuries. Dillenius' work in Oxford made people think about the 'lower plants'. In contrast, Sibthorp's *Flora Graeca* has a mixed legacy. The text reveals Sibthorp's inadequacies, but the quality of the plates, and the completion of the project after nearly six decades, also highlight his ability to pick the right people with whom to work. Druce's comprehensive studies of the flora of Oxfordshire continue to resonate today but reveal the importance of data about the fundamental biology of plants if progress is to be made in understanding plant biology and evolution. By embracing progress in other fields of plant sciences, taxonomic research at Oxford regained its international reputation.

Jacob Bobart the younger (1641–1719)[49] was born in Oxford, the eldest son of Jacob Bobart the elder and his first wife, Mary. Having worked for his father at the Botanic Garden until he was forty, Bobart the younger succeeded to his father's position as superintendent of the Garden. By 1691 he was getting restless in Oxford, and made overtures to the Chelsea Physic Garden, but nothing came of it.[50]

The younger Bobart was widely travelled and highly respected by scholars and gardeners in both Britain and Europe, where he maintained a wide circle of correspondents. One visitor to the Garden thought that his appearance accorded with neither his horticultural nor his academic reputation:

> an unusually pointed and very long nose, small eyes deeply set in his head, a wry mouth with scarcely any upper lip, a large and deep scar on one cheek, and his whole face and hands as black and coarse as those of the meanest gardener or labourer.[51]

As a practical joker, he (in)famously fashioned a rat's corpse into a dragon.[52]

When Robert Morison was killed in 1683, Bobart took on his teaching and academic duties but not his professorial title. In 1699 Bobart completed part 3 of Morison's *Historia*.[53] During the 1680s he fostered a lifelong friendship with the young William Sherard. Months before Bobart's death, Sherard complained about Oxford's treatment of a faithful servant: 'they [the university] ought to have let him spend the short remainder of his time in the Garden'.[54] Bobart the younger, together with his father, is commemorated in the generic name *Bobartia*.

Robert Morison (1620–1683)[55] was born and educated in Aberdeen. In 1644 he was seriously injured fighting for the Royalist cause in Scotland, and escaped to France, where he studied zoology and botany, and finally took a medical degree at Angers in 1648.

Morison came to the attention of the French king's botanist, who recommended him to the household of Gaston, the Duke of Orléans. He worked in the duke's garden at Blois, where he developed his ideas about plant classification and travelled extensively, searching for new species to adorn the ducal garden. It was in Gaston's service that Morison met the future Charles II.

On his restoration to the throne, Charles II made Morison a royal physician and professor of botany. Morison was elected Regius Professor of Botany at Oxford in 1669, the first such position in a British university. The first sign of Morison's dissatisfaction with the prevailing plant classification systems came with the publication of *Praeludia botanica*, which was followed by *Plantarum umbelliferarum distributio nova*, a prospectus for a grand illustrated work, *Plantarum historiae universalis Oxoniensis*.

Morison lived to see the publication of only the second part of this work in 1680. He was killed in a road accident in London. With his death, the university lost an able and popular teacher of botany, whose role was only partially filled by Bobart the younger. Morison was not replaced as professor for over fifty years. He is commemorated in the generic name *Morisonia*, a group of Caribbean members of the caper family.

Johann Dillenius, FRS (1687–1747),[56] was born in Darmstadt, Germany, to a clergyman's daughter and a professor of medicine at the University of Giessen. As a young man, Dillenius practised medicine, but by the late 1710s he was making his reputation as a talented botanist.[57] The publication of his *Catalogus plantarum circa Gissam sponte nascentium* in 1718 brought him to the attention of a botanical patron, William Sherard, who persuaded him to move to England in 1721 and give up medicine. Sherard needed Dillenius' help to arrange his herbarium and compile his *Pinax* of plant names. Sherard's brother James also wanted Dillenius to compile an illustrated catalogue to the rarities growing in his garden at Eltham, Kent.[58] Dillenius devoted himself to his botanical studies.

The work Dillenius completed in London, the third edition of Ray's *Synopsis methodica stirpium Britannicarum* (1724) set a high standard for all the work he published in the remaining twenty-three years of his life. When Sherard finally decided to endow a chair of botany in the university, one of his demands was that the first holder be Dillenius. Dillenius eventually became the first Sherardian Professor of Botany in 1734. He had at his disposal the Botanic Garden, a herbarium he knew intimately, a wide circle of botanical correspondents and a substantial botanical reputation. He made full use of these assets during his tenure, setting a high bar for future holders of the post.

Dillenius died of apoplexy. A modest man, who cared little for the pretensions and trappings associated with his position at Oxford, he was the first and last Sherardian Professor of Botany to publish significant taxonomic research while in post. During his lifetime, Dillenius was honoured by Linnaeus in the name *Dillenia*, a genus of tropical trees, which has 'the showiest flower and fruit, so Dillenius among botanists'.[59]

5016.

1.

2.

W. Fitch del. et lith.

Vincent Brooks Imp.

Elephant apple (*Dillenia indica*), named in honour of Johann Dillenius, from a hand-coloured engraving in *Curtis's Botanical Magazine* (1857). Bodleian Library, Sherardian Library of Plant Taxonomy, t.5016.

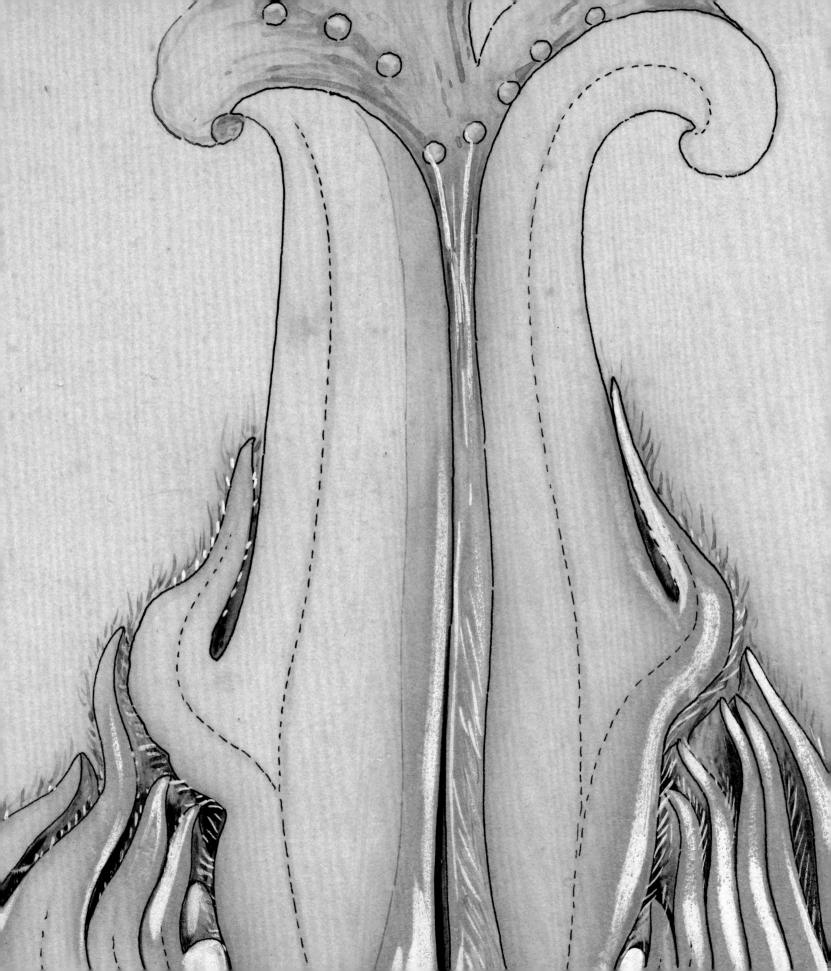

5 FLOWER

Experimental Botany

The sporadic activities that characterized botany at Oxford for over two hundred years from the mid-seventeenth century were typically concerned with cataloguing and classifying plants. As Morison and the Bobarts toiled on their classification, concepts of plant biology were little modified since Aristotle's philosophical speculations. Plants obtained nutrition, which was made in the soil, through their roots, while leaves were little more than protection for young shoots and fruits from the sun and the air.[1] Flowers were understood to give rise to fruits and seeds but their roles in reproduction were ambiguous. Pliny's florid description of fruit production in date palms (*Phoenix dactylifera*) emphasized the need for both male and female trees,[2] although John Parkinson warned his readers, 'I pray you account this among the rest of their fables.'[3] Despite his warning, the case of the date palm was generally accepted but as the exception that proved the rule – plants did not reproduce in the manner of animals. Confusion abounded, with no clear understanding of sexuality and gender, which were rendered the more ambiguous by theological debates about plants and the nature of God.[4]

John Ray critically examined evidence on the morphology, physiology, reproduction, chemistry, ecology, ethnobotany and pathology of plants, something that had not previously been attempted in seventeenth-century Oxford.[5] Ray highlighted where he accepted the facts, where he had doubts and what evidence he incorporated into his classification system. Yet, even at Oxford, approaches to understanding plants, other than through classification, eventually took hold as fundamental questions began to be asked about their biology.

Plant sap

In 1648 John Wilkins, later a brother-in-law of Oliver Cromwell, a founding member of the Royal Society and bishop of Chester, became

Drawing of the male flower of a common oak (*Quercus robur*) made by Arthur Church in 1925 *(detail; see p.157)*

THE BEAR'S EAR

The bear's ear (*Primula auricula*) is a characteristic spring-flowering alpine found in calcareous European mountains. Auriculas were first introduced to British gardeners in the sixteenth century. In 1648 the Botanic Garden was growing a purple bear's ear and a purple-striped bear's ear. By 1658 it boasted nine bear's ears which ranged in colour from tawny through yellow and scarlet to purple and violet.[6]

Jacob Bobart the elder was a well-known auricula breeder, and about a dozen named sorts are preserved in his herbarium. By 1665 bear's ears were described as 'nobler kinds of Cowslips, and now much esteemed, in respect of the many excellent varieties thereof of late years discovered, differing in the size, fashion, and colour of the green leaves, as well as flower'.[7] During the eighteenth and nineteenth centuries, as florists' societies started to take an interest in it, growing auriculas became an obsession for some. Hundreds of sorts were developed, as the upper classes (or more likely their gardeners) and the working classes alike bred these plants and took advantage of two mutations that appeared: a clear green colour and a mealy central ring to the flower.[8]

In the early nineteenth century the botanical writer Robert Thornton admired how the florists – the plant breeders and experimentalists of the eighteenth century – had transformed the wild plants into garden flowers: 'in its wild state it ... attracts no notice from its beauty ... Art accomplishes all the rest.'[9] However, the Swedish botanist Carl Linnaeus was rather less appreciative: 'these men cultivate a science peculiar to themselves, the mysteries of which are known only to the adepts; nor can such knowledge be worth the attention of the botanist; wherefore let no sound botanist ever enter into their societies.'[10]

The Bobarts were also growing other *Primula* species in the seventeenth-century Garden. Cowslip (*P. veris*) and primrose (*P. vulgaris*) were presumably collected locally. The oxlip (*P. elatior*) and bird's-eye primrose (*P. farinosa*) must have been introduced from other parts of England, while *Primula matthioli* would have been introduced from Europe.[11]

Charles Darwin drew scientific attention to the different flower types found within *Primula* species, creating demonstration species that have been used in teaching plant biology across Britain, including Oxford, to the present day.

warden of Wadham College. He was interested in natural philosophy and was a strong advocate of experimental approaches to acquiring knowledge about the natural world.[12] Two men, Robert Hooke and Nehemiah Grew, who were supported by Wilkins, started to change how we view plant function. One such area was the movement of sap through the bodies of plants. From the mid-seventeenth century, botanists searched for systems that might be equivalent to the circulation of the blood in the human body, which had been discovered by William Harvey, warden of Merton College from 1645. At the end of the eighteenth century, John Sibthorp was teaching his students that 'Physiologists hereafter better informed by Experiments will ... make a discovery similar to that of Harvey in the Animal Body of a Circulation'.[13]

Robert Hooke, a chorister at Christ Church in 1653, became Wilkins's assistant. He worked with men in the university, including Robert Boyle and Christopher Wren, and the fledgeling Royal Society to become one of the premier experimental philosophers of the seventeenth century.[14] Hooke was more interested in plants as subjects for microscopy than in understanding them as organisms. His publication of thirteen botanical images in *Micrographia*, the first scientific publication of the Royal Society, showed what the microscope could reveal about the inner workings of plants. Hooke is also credited with the introduction of the word 'cell' into biology.[15]

In contrast to Hooke, the physician Nehemiah Grew was primarily interested in physiology – the ways in which plants grow, feed, move and reproduce. To pursue this Grew had to understand plant anatomy, in which he was actively supported by Wilkins. In 1682 Grew published his magnum opus, *The Anatomy of Plants*. Grew's ideas were based on 'considering that both of them [plants and animals] came at first out of the same *Hand*, and were therefore the *Contrivances* of the same *Wisdom*'.[16] In eighty-two fine copper-plate engravings, he summarized the structure of stems, roots, leaves, flowers, fruits and seeds. His work, together with that of the Italian physician Marcello Malpighi, laid the foundations for the science of plant anatomy.

Grew's intellectual successor was Stephen Hales, who is most famous for being the first person to measure blood pressure.[17] Hales acquired tentative links to Oxford in 1733 with his creation as a doctor of divinity. Around 1715, Hales began to investigate the problem of sap movement through plants; he concluded that the evaporation of water from the

previous page **Auriculas (*Primula auricula*) from Robert Thornton's idiosyncratic *Temple of Flora*** (1807), an attempt to illustrate the principles of Linnaean botany. The aquatint was made by Thomas Sutherland, based on an original oil painting by Philip Reinagle. Bodleian Library, Sherardian Library of Plant Taxonomy, 582 LI 13, A group of auriculas.

opposite ***Vegetable Staticks* (1727), by the Teddington clergyman Stephen Hales**, demonstrated the movement of water from plant roots to their shoots. Over a century later, Oxford botanists showed their first interest in experimental botany. Bodleian Library, Sherardian Library of Plant Taxonomy, Sherard 472, p.28.

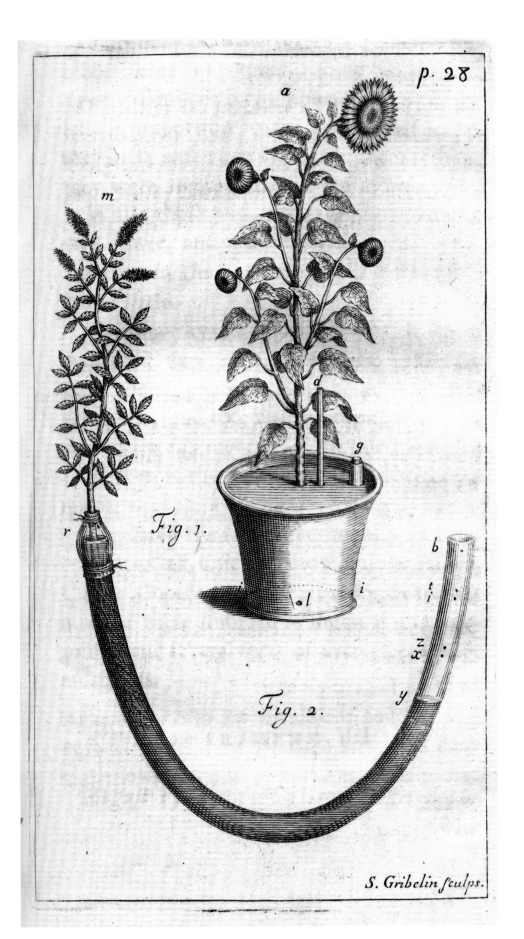

Fig. 1.

Fig. 2.

S. Gribelin sculps.

surface of leaves was sufficient to move water up the plant from the soil. Hales's plant experiments were cleaner and less distressing to his congregation, though no less enlightening, than the vivisection he conducted on horses and dogs. His experimental work is exceptional for the period. His accurate experiments, close, logical reasoning and detailed presentation of data – rather than wild speculation based on limited data and classical authority – are evident in *Vegetable Staticks* (1727). They set a high bar for conducting plant physiology experiments and for reporting results. The father of plant physiology, Hales saw how his science could benefit practical agriculture – a theme that was not taken up again until the start of the nineteenth century.

In contrast, botanical investigations directly connected with the Garden appear to have paid little attention to such developments. The Bobarts focused on practical horticultural skills, such as grafting, where implicit knowledge about sap movement was used daily, or on the accumulation of information about the ways in which plant sap responded to extreme cold.[18] None of these results were presented in the context of wider discussions on plant physiology.

Plant food

Understanding how sap moved around plants was one of numerous plant physiological issues interesting botanists by the end of the eighteenth century. The sources of plant nutrition remained controversial. The Aristotelian view that plant nutrients were made by the soil and then transported into the plant contrasted with the views of experimentalists such as Hales that plants got at least some of their food from the air.[19] Ray observed that light was needed for leaves to become green, although the technological limitations associated with making large pieces of glass contributed to gardeners' persistence in constructing conservatories with tiny windows, including that in the Botanic Garden in the 1670s.[20]

As the Sibthorps occupied the Sherardian chair in the latter half of the eighteenth century, two particularly important experimental groups of observations, which were essential for understanding the nutrition that plants gained from the air, went barely acknowledged by the Garden. Research in the 1770s by the religious dissenter and discoverer of oxygen Joseph Priestley showed that plants demonstrated the effects of 'restoring air which was injured by the burning of candles'.[21] The

Dutch chemist Jan Ingenhousz extended these results, showing that light is essential for the release of oxygen by plants (photosynthesis) and that plants, like animals, also require oxygen for respiration.[22] Leaves are chemical laboratories where air and light combine to make plant food, which is moved around the plant through the sap.

Meanwhile, practical gardeners such as the Bobarts were fully aware of the magic of muck. After all, their domain was built and maintained on the contents of the bladders, bowels, kitchens and stables of Oxford's town and gown.[23] Their forays into the subject of soil-based plant nutrition were associated with the practical benefits of different types of manure; the well-rotted contents of New College's 'house of office' was the preferred vine manure in the Botanic Garden.[24] Moreover, in an age of widespread alchemical belief, the Bobarts were convinced that the soil of the Botanic Garden could transform 'Crocus' into 'Gladiolus' and 'Leucoium' into 'Hyacinth'.[25] Of more interest is Robert Morison's fascinating discussion of different sorts of brassica, and how he believed they changed from one sort into another in different soils and in different places.[26]

The chemistry of manures was taken up by Humphry Davy in his *Elements of Agricultural Chemistry* (1813), but it took the election of Charles Daubeny to the Sherardian chair in 1834 for experimental interest in soil nutrients to become part of botanical research at Oxford (see below). Unlike most of his predecessors and successors, Daubeny did his most important botanical research while he was Sherardian professor. In an influential series of experiments, he demonstrated the effects of different coloured lights on the production of oxygen in a wide variety of plants, presumably taken from the Garden.[27] The different colours were produced by passing light through pieces of coloured glass and solutions of coloured chemicals: port wine (red) and a solution of copper sulphate and ammonium (blue-violet).

By the end of the nineteenth century, that green plants obtained mineral nutrients and water from the soil, and organic compounds through photosynthesis, was well known. The discovery of the fundamentals of these processes would take most of the twentieth century, and involve contributions from botanists at Oxford. The details of picking apart the cellular, biochemical and molecular bases of these processes are major occupations of research groups in the current Department of Plant Sciences.

A footnote in the history of plant nutrition is provided by the research of Sydney Vines, started when he was an assistant at the newly opened Jodrell Laboratory at Kew in 1876 and continued at Oxford in the 1890s. Vines demonstrated the presence of enzymes that could break down protein in the pitchers of the carnivorous genus *Nepenthes*.[28]

Daubeny and plant nutrition

Within the University of Oxford, and its sister institution at Cambridge, there was little concern with knowledge that might have practical value, except medicine and law, until the latter half of the nineteenth century.[29] Towards the end of the eighteenth century, John Sibthorp had directed his lectures at men who were interested in agriculture but had moved to teaching academic botanical rather than technical knowledge. Rudimentary investigations of plant physiology in the university suffered similar drawbacks. The application of manure was the province of horticulturists, a practical task rather than one involving academic consideration of how fertilizers affect individual plant species.[30] Charles Daubeny, that 'zealous propagator of scientific principles in agriculture',[31] tried to change this and to establish agriculture as a scientific discipline in Britain.

Daubeny already held a prestigious, if poorly paid, position as professor of chemistry, and had a distinguished record of research and teaching in geology and chemistry, when he was elected the fifth Sherardian professor in 1834. When he became the first Sibthorpian Professor of Rural Economy in 1840 he had space at the Botanic Garden, personal wealth and the skills to address questions about plant nutrition that had vexed generations of botanists. He also went on to build his own chemical laboratory just outside the Botanic Garden.[32]

Using controlled field trials, initially set up in the Garden, Daubeny began to explore why crops grown continuously on the same piece of land show yield reductions over time. This problem, known to farmers for thousands of years, was traditionally overcome by crop rotation. Daubeny rejected the traditional notion that this was caused by toxins secreted by plants, which gradually accumulated in the soil. By measuring yield changes and the mineral content of the soil, he revealed that plants remove essential nutrients from the soil as they grow,[33] and that some nutrients in the soil are available to plants, while others remain unavailable. Daubeny was a strong advocate of the

Mid-nineteenth-century oil portrait of Charles Daubeny, the fifth Sherardian Professor of Botany, who revived the fortunes of the Botanic Garden. University of Oxford, Department of Plant Sciences.

German chemist Justus von Liebig's views that the organic components of the soil matter little to plant nutrition, and that what is important is the availability of minerals. That is, if enough minerals are available in the soil, plants obtain as much nitrogen as they need from the air. At this time, large-scale field trials at Rothamsted Manor (later Rothamsted Experimental Station) by the agricultural scientist and businessman John Bennet Lawes were starting to show the importance of nitrogenous fertilizers for crop production.[34] Daubeny was among the few nineteenth-century agricultural researchers working outside Rothamsted or the agricultural societies that had emerged in the previous century.[35]

It is unclear whether Daubeny influenced Lawes's ideas about agricultural fertilizers.[36] Lawes matriculated in 1833 and signed a register of those attending Daubeny's lecture in 1835, when, according to his own accounts, he was already experimenting with manures, using superphosphates, and with plants in pots as early as 1834.[37] Lawes himself stated that 'Eton and Oxford were not of much assistance to those whose tastes were scientific rather than classical, and consequently my early pursuits were of a most desultory character'.[38]

Daubeny's experimentation contributed to our knowledge that plants need nutrients such as nitrogen, phosphorus and potassium, and that these nutrients come from the soil. The fundamental role of the interactions between soils, beneficial microbes and the atmosphere in the maintenance of soil productivity was taken up briefly by Sydney Vines in the late 1880s.[39] The enhancement of plant productivity through experimental work on soil microbes and atmospheric gases is an important strand of research in the Department of Plant Sciences in Oxford today.

Daubeny laid the foundation for scientific agriculture in the university but that foundation crumbled after his death, only to be re-established in the early twentieth century. The collegiate university indirectly benefited from agricultural chemistry in the nineteenth century. In Daubeny's time a major source of nitrogen and phosphorus for soil improvement was South American guano. The chief guano trader was Antony Gibbs & Sons, of which William Gibbs, the funder of Keble College Chapel, was a founding partner.[40]

A thin section of a fossil, made in the late nineteenth century, which proved fundamental to understanding how the roots of extinct fossil plants grew nearly 400 million years ago. Oxford University Herbaria, Foss-slide_081.

Growing plants

By the late seventeenth century, the Bobarts had learned to grow many unusual plants in Oxford, including sensitive plants, which droop when touched. Through trial and error, gardeners knew that plants are affected by their environments, and made enormous efforts to create conditions that would enable exotic plants to grow. They were familiar with the phenomenon whereby the roots of germinated seeds grow down and the shoots grow up, no matter which way they are planted. One explanation in circulation by the end of the eighteenth century was that roots responded to gravity.[41] In the early nineteenth century, Thomas Knight decided to investigate some of these ideas.

In 1788 Thomas Andrew Knight entered Balliol College but, like many young men of his class in eighteenth-century Oxford, he never graduated.[42] Inherited wealth and a family estate enabled him to pursue investigations into plant physiology that were initially suggested to him by Joseph Banks.[43] With great mechanical skill and a flair for inventing novel pieces of scientific apparatus, Knight investigated many of the physiological questions of his day, seeking mechanical explanations for plant movement in relation to environmental factors.

In the 1920s Frederick Keeble, the ninth Sherardian Professor of Botany and, like Knight, a keen horticulturalist, encouraged the young botanist George Robert Sabine Snow to investigate the sensitivity of plants to external stimuli.[44] From the 1920s George Snow and his wife Mary (née Pilkington) collaborated on experimental work that changed our understanding of how plants respond to gravity. Mary Pilkington graduated from St Hugh's College in 1926 and became Snow's first research student. The facilities offered by the university for botanical research were poor, even when only minimal amounts of equipment were necessary. The Snows took the nineteenth-century option and converted part of their own home and garden in Headington into laboratories and plant-growing facilities. Neither George nor Mary, who held positions at Magdalen College and Somerville College, respectively, were ever formally members of the Department of Botany.

The Snows' approach was to conduct carefully designed experiments that answered single questions unambiguously. Experiments that relied on a minimal amount of specifically designed equipment and immense technical skill were conducted on living plants. The Snows' scientific collaboration proved very productive. They designed new experiments

Drawing of the male flower of a common oak (*Quercus robur*) made by Arthur Church in 1925 to illustrate teaching on floral structure and phyllotaxy. Bodleian Library, Sherardian Library of Plant Taxonomy, MS. Sherard 406, f.8r.

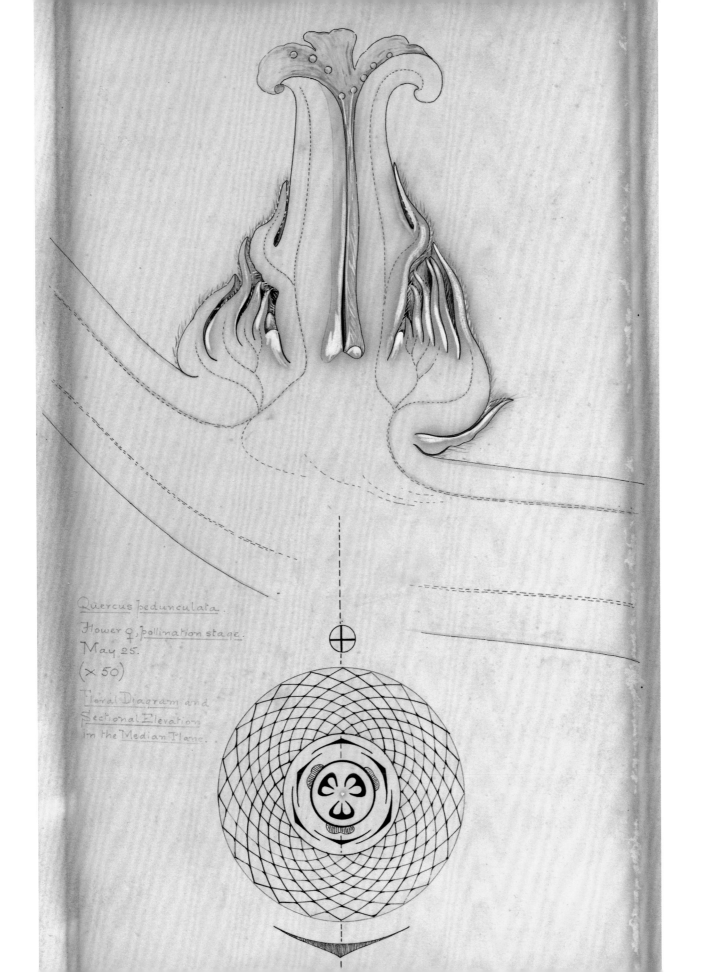

Quercus pedunculata.

Flower ♀, pollination stage.

May 25.

(× 50)

Floral Diagram and
Sectional Elevation
in the Median Plane.

to test their ideas based on their experience with previous experiments. Results from the experiments, almost all of which were undertaken by Mary, were jointly interpreted, written up by George and eventually published under their joint names.

External stimuli and plants were not the only aspects of plant biology that interested the Snows. They were interested in the processes that are responsible for the arrangement of plant leaves (phyllotaxis) during development. Since the early 1900s, this age-old problem had received theoretical attention from one of George Snow's teachers, Arthur Harry Church.[45] Experimental data about phyllotaxis were gathered by the steady hands of Mary Snow and contributed to papers that are still cited, more than eighty years after their publication, in cutting-edge plant developmental biology research undertaken at the Department of Plant Sciences today.

Inside the cell

By the end of the nineteenth century, light microscopy was well established as a research and teaching tool in British botanical institutions. At Oxford the focus was on anatomy and embryology, particularly of ferns and liverworts, in the work of John Bretland Farmer. Farmer was appointed demonstrator of botany by Isaac Balfour in 1887, before moving to Imperial College in 1892, where he eventually became professor of botany. Farmer is credited as one of the originators of the term 'meiosis'.[46]

Serendipity and the prepared mind are important elements of research. In the early 1950s experimental botany at Oxford was buoyant. Botany had moved into new, purpose-built facilities on South Parks Road, while the recently elected twelfth Sherardian Professor of Botany, the flamboyant and dynamic Cyril Darlington, was starting to recruit new talent, with diverse academic interests, to his department.[47] Darlington encouraged his staff to pursue research in genetics, structural biology of the cell and, of course, chromosomes, although his interest in other areas of botanical research was limited.

Lionel Frederick Albert Clowes completed his doctorate on the anatomy of beech tree roots, using classical microscopical approaches, at Magdalen College in 1949.[48] His discovery of clusters of apparently inactive cells (later dubbed 'the quiescent centre') at the tips of these roots occupied the rest of his career. Clowes's initial challenge was to convince other scientists that the cells were truly inactive. The Snows' surgical techniques were too crude for minute roots, so Clowes used photographic film to detect where radioactively labelled chemicals accumulated, indicating active cells. He obtained the proof he needed, and went on to demonstrate the generality of the quiescent centre across most land plants.[49] In his *Apical meristems* (1961), Clowes synthesized his ideas about the growing points of plants, while *Plant Cells* (1968), co-authored with Barrie Juniper, a young electron microscopist in the department, explored the insides of plant cells.

Darlington's discipline of cytogenetics, the fusion of cytology and genetics, which had been part of research programmes in forward-looking botany departments across Britain and internationally since the 1930s, did not find a foothold at Oxford. Darlington and his students soon started research programmes on the taxonomic distribution of chromosome numbers within and between species, and on the behaviour of chromosomes within cells.[50] In the late 1950s Darlington's doctoral student Douglas Davidson began using X-rays to investigate their effects on chromosomes. Clowes used X-rays and similar approaches to demonstrate that the quiescent centre was a reservoir of cells from which root growth could be restored after stress. In further studies, Clowes determined the rates of cell division, and the patterns of behaviour of cells, in the growing tips of roots.

The ideas spawned by the Snows and by Clowes, which derived from rigorous experimentation at Oxford from the 1920s into the 1970s, are fundamental to plant developmental research. They have now become part of 'standard' botanical knowledge, and so are rarely even attributed – perhaps the ultimate measure of scientific achievement.

Function of the flower

The idea that plants reproduce sexually is taken for granted, but in the eighteenth century the idea was sufficiently modern to be outrageous. In *The Anatomy of Plants* (1682) Nehemiah Grew presented the structure of the flower in terms of the 'empalement' (calyx), the 'foliation' (corolla) and the 'attire' (everything inside the corolla). Within Grew's confusing explanation of the function of the 'attire', drawing as it does analogies from animal reproduction, is the statement: 'In discourse hereof with our Learned Savilian [*sic*] Professor Sir *Thomas Millington*, he told me, he conceived, That the *Attire* doth serve, as the Male, for the *Generation* of the Seed. I immediately reply'd, That I was of the same Opinion.'[51] Millington's, and hence Oxford's, claim of fundamental involvement in the discovery of plant sexuality has been a cause of academic debate ever since.[52]

Whatever the claims and counter-claims, unpicking the functions of flower parts, together with the mechanism and biological significance of plant sex, took place outside Oxford. By the end of the eighteenth century, three German researchers had established the essentials of flowering plant reproduction.[53] Rudolf Jakob Camerer, at the University of Tübingen, Germany, produced the first experimental evidence for plant sexual reproduction in *De sexu plantarum epistola* (1694). Between 1761 and 1766 Joseph Gottlieb Kölreuter, at the University of Karlsruhe, Germany, published seminal papers on crossing within and between plant species. Christian Konrad Sprengel's *Das entdeckte Geheimnis der Natur im Bau und in der Befruchtung der Blumen* (1793) demonstrated the intimate association between flowers and their insect visitors. In 1717 the French botanist Sebastien Vaillant published a paper on plant sexuality that influenced the work of Linnaeus and the creation of his controversial sexual system of classification.[54] The significance of Kölreuter's and Sprengel's work, and the biological importance of plant sex, did not become clear until the work of Charles Darwin in the 1860s and 1870s.

The implications of these ideas were overlooked at Oxford, as elsewhere. The nineteenth-century German botanist Carl Friedrich von Gärtner offered a partial explanation as to why academic communities resisted investigation of the function of plant sexuality: 'they [hybrids] were attacked to such a degree that their genuineness was doubted and strenuously contradicted, or else they were regarded as a sort of inoculation phenomenon belonging to gardening.'[55]

Some observations were being made at Oxford that were relevant to plant sex and hybridization, but the conclusions were never formalized. Bobart the younger found a white campion with flowers that lacked male parts, and he was aware, like many gardeners, that plants such as cannabis had individuals that did and did not produce seed.[56] Moreover, before 1674, he had told John Ray that he had raised primroses and oxlips from cowslips.[57] Bobart the younger is also credited with the recognition of the hybrid tree, the London plane (*Platanus* x *hispanica*), which he described as intermediate between the occidental

Frontispiece of Sprengel's *Das entdeckte Geheimnis der Natur im Bau und in der Befruchtung der Blumen* (1793), the first book to demonstrate the importance of insects in floral biology. Bodleian Library, Sherardian Library of Plant Taxonomy, 581 SP1.

Das

entdeckte Geheimniss

der

NATUR

im Bau und in der Befruchtung

der

Blumen

von

CHRISTIAN KONRAD SPRENGEL,

Mit 25 Kupfertafeln.

Berlin, 1793.

bei Friedrich Vieweg dem ältern.

C. Jäck Scripsit et Sculpsit

Gezeichnet v. C. K. Sprengel.   W. Arndt Sculp:

mr Fairchilds
mula

Herb. Sherard

569

(*P. occidentalis*) and oriental (*P. orientalis*) planes.[58] Among the specimens in Sherard's herbarium is one of only two known examples of the first artificially created hybrid plant, Fairchild's mule. The mule, a hybrid between a carnation and a sweet william, was made by the nurseryman Thomas Fairchild of Hoxton in about 1717, and brought to academic attention by the horticulturalist Richard Bradley, who was to become the first professor of botany at Cambridge.[59] At Oxford, the significance of this specimen of Fairchild's mule, and what it represented for the evolution of plant species and plant breeding, went unremarked until the late twentieth century.[60] Elsewhere, the possibility that hybrids occurred naturally led people to start questioning, at least implicitly, the assumption that species numbers had been fixed at the Creation.

Practical horticulturalists were starting to make crosses within and between plant species to improve crops or garden plants. Gardens, nurseries and orchards provided an ideal location to put ideas of plant improvement into practice; they were, and are, a plant hybridizer's paradise. However, these plant improvers were often working in isolation from each other and from the wider scientific community. Prominent among the early plant improvers was Thomas Knight. Working at the start of the nineteenth century, Knight was pragmatic enough to realize that if he wanted quick results he needed a plant with a short life cycle; the plant he chose was the annual garden pea. Through hybridizations between different types of garden pea (*Pisum sativum*), Knight made considerable progress in showing how variation was inherited from one generation to the next. However, unlike Gregor Mendel who chose to cross peas later in the same century, Knight did not count the different offspring that resulted from the crosses.[61]

By the end of the nineteenth century, flower structure and function had become central to botanical teaching and research. In early twentieth-century Oxford, Arthur Harry Church began detailed investigations of floral morphology, culminating in his *Types of Floral Mechanism* (1908), accompanied by his immaculate botanical illustrations.[62] At Oxford, the establishment of research programmes investigating the details of reproductive barriers between species had to await the genetic, biochemical and molecular genetic techniques that were introduced towards the end of the century. Today, information derived from such techniques is used by almost all research programmes within the Department of Plant Sciences.

Fairchild's mule, the first artificially created plant hybrid, was made by the gardener Thomas Fairchild by crossing a carnation (*Dianthus caryophyllus*) and a sweet william (*Dianthus barbatus*) c.1717, at his nursery in Hoxton. Oxford University Herbaria, Sher-0569-50.

Ecology

Until the mid-twentieth century, genetic studies were not a major part of botanical research programmes at Oxford, but took hold elsewhere in the university. By the 1930s Eric Brisco Ford had firmly established genetic studies in the Department of Zoology. Ford, with his interests in genetics and field natural history, created the discipline of ecological genetics, where ecology and population genetics overlap, and the evolutionary consequences of reproduction are investigated,[63] though many of his findings have been questioned. In the Department of Botany, such interests would not be reflected in the research interests of academic staff until the 1960s.

With the arrival of Arthur Tansley as the tenth Sherardian professor in 1927, the Botany Department at Oxford strengthened its interests in plant ecology, which had begun with Church in the early part of the century.[64] Tansley arrived from the University of Cambridge as the foremost ecologist in the British Isles who was active as both teacher and researcher.[65] Like Vines, Tansley found Oxford a difficult place in which to undertake research, though he recruited some talented young staff to the department, such as Arthur Clapham, and nurtured their successful academic careers. Perhaps most significantly, he brought to Oxford the editorship of the *New Phytologist* and the *Journal of Ecology*, both of which he had founded and which were two of the foremost journals for publishing ecological research papers.

When Tansley retired in 1937 he ensured that his successor would be an ecologist. The choice of the electors as eleventh Sherardian professor was Theodore George Bentley Osborn, who at the age of twenty-five had established the Department of Botany at the University of Adelaide.[66] Osborn arrived at Oxford with wide ecological interests, especially in Australian vegetation. Tansley had made few changes to the physical structure of the Department of Botany. This was left to Osborn, who was the architect of the move that split the Department of Botany from the Garden. The split had dramatic effects:

> not only was the production of original work in botany greater in quantity and quality than ever before but of the undergraduates, demonstrators and research men present in Oxford during his tenure of the chair, 11 or more are or have been professors and heads of departments, and several more Readers in biological departments of universities in Britain and elsewhere.[67]

Arthur Harry Church's section through a flower of the snowdrop *Galanthus elwesii* as published in his *Types of Floral Mechanism* (1908). Bodleian Library, Sherardian Library of Plant Taxonomy, 581.1 CHU /BT(OS), p.31.

A crowded Plant Physiology Laboratory at the Botanic Garden in the 1930s before the Department of Botany moved to its present site on South Parks Road. Oxford University Herbaria.

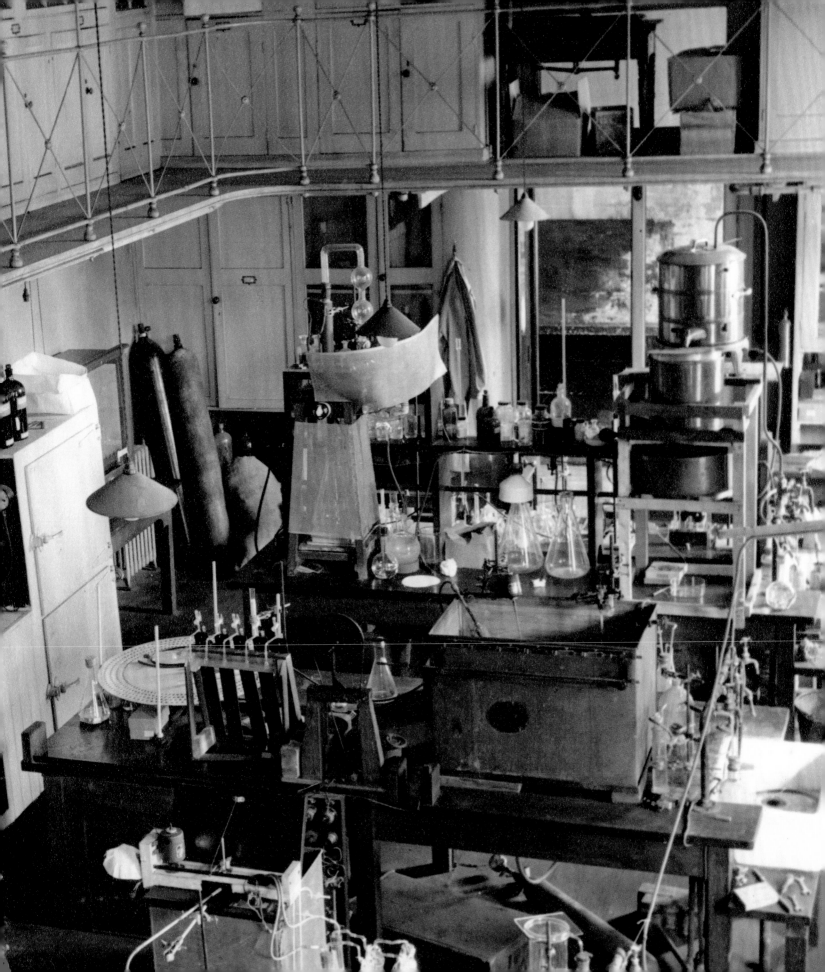

Among these was Nicholas Polunin, whose doctoral research on the ecology of Arctic plants had been started under Tansley, but who maintained a post in department until after the Second World War.[68]

Despite Darlington's antipathy towards ecology, ecological research continued in the Department of Botany through the activities of researchers such as Stan Woodell. However, it was in the Departments of Forestry and Agriculture where ecological research on plants developed most strongly. For example, Frank White and his colleagues and students in forestry mapped the African vegetation.[69] Plant sciences at Oxford remain strong in both pure and applied ecology.

Changes in the technology of molecular biology, together with ready access to computing power and near-instantaneous communication across the globe, have dramatically altered the way in which plant sciences are done at Oxford, as elsewhere. Today research is rarely undertaken by a single scholar; academics are surrounded by a team of research students and postdoctoral research assistants, and often work with colleagues across the world. Consequently, the questions asked, the hypotheses tested and the answers generated are the product of collaboration. In addition to the technical and intellectual climates in which modern plant sciences researchers work, the political climate has changed. Researchers must be willing to explain their work to many different audiences, which, in a different age, would have been alien.

Cyril Dean Darlington, FRS (1903–1981),[70] was
a cytogeneticist who became twelfth Sherardian Professor
of Botany. Born in Lancashire, Darlington took a degree in
agriculture from London in 1923. Starting as a volunteer in
the Cytology Department of the John Innes Horticultural
Institution in London, he eventually gained the position of
cytologist, then ran the department until, at the age of thirty-
six, he became director of the institution.

A largely self-taught cytologist and geneticist, Darlington
synthesized the two fields into a new discipline of his own
creation – cytogenetics – in two books, *Recent Advances in
Cytology* (1932) and *The Evolution of Genetic Systems* (1939).
These books attracted researchers to the institution, and
created a world-leading centre for cytogenetics. Darlington's
insights into the behaviour of chromosomes are critical to our
understanding of the mechanisms of plant evolution.

Darlington was elected to the Sherardian chair in 1953.
Opinionated and iconoclastic, he frequently set himself
against establishments, so his arrival led to inevitable ructions
among the existing staff.[71] The cytogenetics that had made his
reputation was being eclipsed by new methods and insights to
be gained from molecular biology. Furthermore, his interests
were moving towards the interface of genetics, humans and
society, culminating in the highly controversial *Evolution of Man
and Society* (1969).

During the eighteen years he was Sherardian professor,
Darlington established the Botany School in its new home
away from the Botanic Garden. As an academic committed
to research, he fought to establish the Harcourt Arboretum,
recognizing that areas where large-scale experiments could be
conducted would be critical for future plant sciences research
in the university.

Sydney Howard Vines, FRS (1849–1934),[72] the only child of an Ealing businessman and his wife, spent his early life on a Paraguayan sheep ranch. Disillusioned by three years of medical training, Vines won a scholarship to Cambridge, where he headed the results list for natural sciences in 1875.

Over the next decade, translating, teaching, writing and some research took over Vines's life. Following a short period at the Royal School of Mines and the Royal Botanic Garden, Kew, Vines established himself in Cambridge, where he transformed botanical teaching. Fluent in German, Vines worked with scientists in Germany (e.g. Julius von Sachs and Heinrich de Bary, the founder of plant pathology), forging a new approach to botany in late nineteenth-century Europe. Vines's translations of Karl Prantl's *Lehrbuch der Botanik* (*Elementary Botany*, 1881) and Sachs's *Lehrbuch der Botanik* (*Textbook of Botany*, 1882) brought German botany to an English-speaking audience. In 1887, the year before his election as the eighth Sherardian Professor of Botany, Vines helped found the *Annals of Botany*; he was the journal's editor until 1899.[73]

Vines failed to repeat in Oxford the transformations he had wrought at Cambridge. Indeed, rather than attracting undergraduates to botany, his reputation was putting them off, and he showed an even greater degree of prejudice against women than was typical at the time at Oxford.[74] Yet during his presidency of the Linnean Society, he supported the decision in 1903 that women and men be admitted to fellowships on equal terms.

Christine Mary Snow (1902–1978)[75] was the daughter of Alfred Pilkington, of the glass manufacturer Pilkington Brothers. She joined St Hugh's College, Oxford, when she was twenty, and took a first-class degree in botany in 1926, before becoming the first research student of George Snow, a botanist and fellow of Magdalen College. Her marriage to Snow in 1930 prevented her taking up a research fellowship to which she had been elected by Somerville College.

The financial independence of both Mary and her husband meant that they could abandon the limited facilities offered by the university for botanical research to create their own facility at their home in Headington. Here the Snows entertained and taught students and researchers, as they focused on intricate experimental work aimed at understanding plant interactions with the environment and plant development. The surgical procedures needed to manipulate the growing tips of plants, which characterized the Snows' research, were primarily carried out by Mary. From the 1930s to the early 1960s, flurries of papers were published jointly by Mary, with her husband as junior author, on aspects of leaf development.

As a practising experimentalist, Mary knew that botany needed facilities and that such facilities could be costly. Consequently, she became a benefactress to both the Botanic Garden and the Department of Botany, following its move to South Parks Road in the early 1950s. Today, an annual lecture in the Department of Plant Sciences bears her name.

6 FRUIT

Applied Botany

Since antiquity, medicine and food have been used to justify the study of plants. The Botanic Garden was founded on the utilitarian value of plants as medicine. Botanic gardens around the world were established with the prospect of economic power fuelled by plants. For example, the Royal Botanic Gardens, Kew, was at the centre of a global network of colonial gardens established to realize the capital value of controlling economically important plants.[1] Evidence of Britain's wealth over the centuries could be seen in the buildings of Oxford University and the many endowments created to sustain it.[2] Much of this wealth was acquired directly or indirectly though trade in natural products, especially plants. Plants are also at the very heart of the university: they formed the paper and ink upon which data, information and ideas have been passed from generation to generation.

Gentlemen who studied at the university, and many of the academics, acquired their wealth and incomes, as landowners, physicians or prelates, from agriculture. The third Sherardian professor, John Sibthorp, taught future landowners the rudiments of botany, but his main income came from agriculture on estates inherited from his mother.[3] His father, Humphrey, survived his long academic career at Oxford on the income from the Sibthorp family estate in Lincolnshire.[4] Income from John Sibthorp's estate at Sutton and Stanton Harcourt, approximately ten kilometres west of Oxford, subsidized the production of the *Flora Graeca*, and initially contributed towards the stipend for the Sibthorpian Chair of Rural Economy.

Despite its dependence on plants, the university was reluctant to wholly embrace the more applied aspects of botany and the role of plants in society until the twentieth century. Forestry and agriculture, the two most obvious applications of plant sciences (other than as physic), naturally demand collaboration across diverse disciplines including botany, zoology, geology, economics, sociology and politics – a diversity that is difficult to maintain within academic departments in a traditional university.

Watercolour of black pine (*Pinus nigra*) by Rosemary Wise *(detail; see p.177)*

THE YEW

Yew (*Taxus baccata*)[5] is a conifer that is native to Europe, North Africa and south-western Asia. In the 1300s yew was the preferred raw material for making high-quality bows. Yew wood staves, cut from the junction of the heartwood and the sapwood, have an ideal combination of properties for bow-makers to fashion deadly weapons. As wars that relied on bowmen were fought, yew trees were harvested across Europe to supply an extensive trade, and laws were created to protect supplies.

By the seventeenth century, yew was a favoured hedging plant, with gardeners asserting their power over nature through the art of topiary. The Botanic Garden was no exception, as Bobart the elder laid out geometric hedges and clipped two yews to be 'Gigantick bulkey fellows, one holding a Bill th' other a Club on his shoulder'.[6] A yew from one of Bobart's hedges survives in the Garden, which, spared the shears in the mid-nineteenth century, became a tree.[7]

In the late twentieth century the value of yew changed yet again. Yew is well known as a poisonous plant, but plants that are toxic at a particular dose may have medicinal potential at a lower dose.

Consequently, yew extracts, which were used for thousands of years in traditional healing systems, produced orthodox medicine's best-selling anti-cancer drug – paclitaxel. First isolated from the bark of a species of North American yew in the mid-1960s, paclitaxel was given approval for use in chemotherapy in the early 1990s.

The hope it offered, and the accompanying hype, meant that yew once again became a strategic resource, with yew bark in great demand. Yew populations cannot sustain the demand for thousands of tonnes of bark per year. The problem is that paclitaxel has a complex structure that it is uneconomic to synthesize artificially. Eventually a compound was discovered in yew foliage that could be easily and economically converted to paclitaxel. Yew-hedge clippings became the means to satisfy the ever-increasing demand for the drug.

The changing ways in which we use yew have seen the tree disappear from native forests across Europe, though it is commonplace in gardens and urban landscapes. The conservation of native yews requires biological research, integrated with research from across many other disciplines.

1.   3.   4.   5.   6.   2.

Respectability and applied research

Until the mid-nineteenth century, Sherardian professors had private incomes or could maintain their lifestyle from the stipend associated with the post, fees from teaching and the benefits of a college fellowship. Unusually, Daubeny resigned his position as a physician at the Radcliffe Infirmary in 1829, which he had used to support his stipend as professor of chemistry, to become a professional scientist, apparently on the basis of knowledge about how the university would bestow the botanical chairs it had in its gift.[8] Five years later he became Sherardian professor, and then, in 1840, the Sibthorpian professor.

The tension between being a gentleman of science, who did not get paid for doing science, and the need for an income on which to live was replicated across the university and British society more generally.[9] A typical nineteenth-century Oxford education was based on a curriculum for the moral and intellectual development of undergraduates to prepare them for an 'active and purposeful life'[10] beyond the city's spires. Its defenders even suggested that the sciences might be better suited to the education of experts than to the general education of gentlemen.

The idea that gentlemen could become professional scientists was resisted for much of the nineteenth century, but broadly applied subject areas for teaching and research were popping up across Oxford. These were often promoted in manners designed not to offend the strong artistic and theological sensibilities of the university. By the end of the nineteenth century, professional science had become established in the United Kingdom as an acceptable pursuit of gentlemen. Between the world wars, science as a profession and the teaching of applied subjects within the university gained a degree of respectability. In botany, the applied subjects were agriculture and forestry, subjects that had attracted the attention of Daubeny.

Agriculture

The university had decided that the holder of the Sherardian chair would also hold the Sibthorpian Chair of Rural Economy. Independent of this decision, when funds from Sibthorp's estate became available in 1840, the obvious choice was Daubeny. He had demonstrated his commitment to the Garden, and his interests in soil chemistry were ideal for a post that was primarily concerned with agricultural production. Daubeny's duty as Sibthorpian professor was to read a public lecture on rural

previous page **Teaching wall chart, published as part of Leopold Kny's series *Botanische Wandtafeln*** (1874–1911), of yew (*Taxus baccata*), the oldest tree in the modern Botanic Garden and a modern living link with Bobart the elder. Oxford University Herbaria.

opposite **Watercolour of black pine (*Pinus nigra*)**, made by Rosemary Wise in 2013, from a specimen planted by Charles Daubeny in the 1830s and felled in 2015. Oxford University Herbaria.

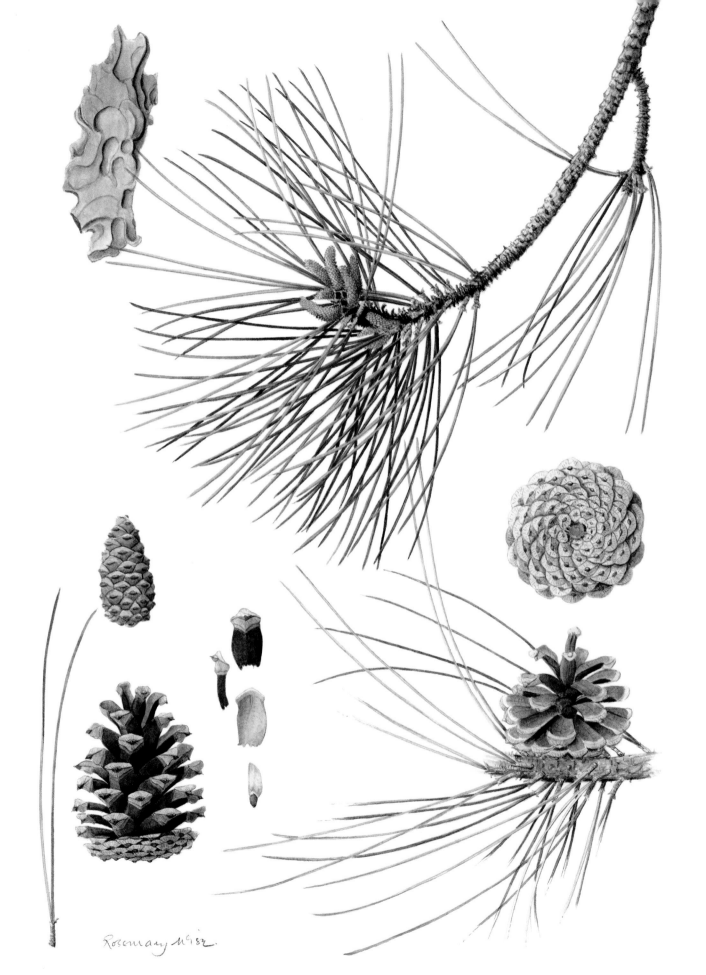

Rosemary M[^i]82.

economy each term. By 1841 he was advocating a formal system of agricultural education, principles that would be adopted when the Royal Agricultural College, Cirencester, was established in 1845 by a committee headed by Henry Bathurst, fourth Earl Bathurst, a graduate of Christ Church.[11]

In 1882 the university separated the Sherardian and Sibthorpian professorships; Marmaduke Alexander Lawson was the last holder of both positions. However, the value of Sibthorp's endowment had diminished so much that a 'visiting professorship' was established, whose holder was to give twelve lectures on agricultural subjects. The first man in this office, Joseph Henry Gilbert, director of the chemical laboratory at Rothamsted,[12] held the position between 1884 and 1890. His qualifications, through his collaborations with John Bennet Lawes on applying chemistry and plant physiology to practical agriculture, were impeccable.[13] The next holder of the post, between 1894 and 1897, Robert Warington, was also a distinguished agricultural chemist, another product of Rothamsted.[14] By the end of the century Sydney Vines was fighting the cause of botany against the threat he perceived from agriculture.[15]

In 1907, with financial help from St John's College, a full-time Sibthorpian professor was appointed, and a School of Rural Economy created in buildings erected on Parks Road. The Scottish agriculturalist William Somerville occupied the post until 1925.[16] Somerville had been the first lecturer of forestry in Edinburgh University, then professor of agriculture and forestry at Durham College of Science (later Newcastle University) and the first professor of agriculture at Cambridge University, before arriving at Oxford. In addition to his substantial contributions to agricultural research, Somerville was recognized as a gifted communicator. The manner in which he laid out experimental plots, as visual demonstrations, showed what science could do for plant growth and helped reduce suspicions about agricultural research among farmers.[17] Under Somerville's stewardship, the school offered a diploma in rural economy. By the end of the First World War students could also study for a bachelor's degree.

When in 1925 the Scottish agriculturalist James Anderson Scott Watson became Sibthorpian professor, a post he held for twenty years, there were few high-quality students and the school faced decline.[18] Like Vines, Watson struggled to repeat his previous achievements at Oxford.

However, by 1937 the Ministry of Agriculture had approved agriculture becoming an honours degree at Oxford. The syllabus of agricultural science and economics was augmented with the investigation of the evolution of agriculture, a subject of special interest to Watson as it had been to Daubeny.[19]

The end of the Second World War saw the appointment of Geoffrey Emett Blackman, who held the post until 1970, and the school become the Department of Agriculture. Nomenclature was not the only thing to change under Blackman. The new professor brought statistical rigour to botanical research at Oxford and shifted agriculture teaching from farm management back to agricultural science. The result was more staff and

Photograph by Arthur Church of oat fields around Oxford in the early twentieth century. Oxford University Herbaria.

students, more space and the establishment of the university field station at Wytham in 1956. Applied biological sciences became a full part of the academic community in the university.

The 1970s saw the British mycologist John Harrison Burnett in the professorial position. The Department of Agriculture consolidated, taking advantage of its status, and a new three-year honours degree in agricultural and forest sciences was established. However, by the early 1980s, as the tenure of David Cecil Smith, an authority on plant symbiosis, began, the national situation had changed.[20]

In 1981 the British government withdrew funding for all teaching in agriculture and forestry at Oxford. The result was the creation of a new degree in pure and applied biology and the closure of the Department of Agriculture. In 1985 the Department of Plant Sciences was created by merging the Departments of Agriculture, Botany and Forestry. In 1989 the Sibthorpian Professor of Rural Economy was rechristened the Sibthorpian Professor of Plant Sciences, and in 1990 the plant biochemist Christopher John Leaver, from the University of Edinburgh, took up the post and leadership of the Department of Plant Sciences.

Forestry

Universities and other organizations across the United Kingdom were suspicious of forestry as a discipline, yet by the 1870s people who knew about the practical aspects of tree management were in short supply across the British empire, especially in India.[21] Meanwhile, forest education in nineteenth-century Europe, especially Germany and France, was showing the economic value of applying science to forestry and creating a well-trained workforce. It was not until 1885 that England had its first professor of forestry, William Schlich. An eminent German forester with a wealth of experience in Indian forestry, Schlich took up the post at the Royal Indian Engineering College at Cooper's Hill, near Egham, Surrey. The many editions of his *Manual of Forestry* became the standard forestry textbook, while his influence shaped the establishment of the British Forestry Commission in 1919. When he retired in 1920 (at eighty years of age), Schlich's ideas had permeated forestry throughout the British empire.

In 1905 Cooper's Hill closed, and Schlich moved the Imperial Forest School, lock, stock and barrel, to Oxford. By 1908 St John's College had extended the buildings occupied by the Sibthorpian professor. St John's also provided Schlich with Bagley Wood, a piece of woodland some eight kilometres south-west of Oxford that has survived since before 1600, as a teaching facility. Initially, students studied for a two-year diploma, without any prior qualification, but by 1909 an undergraduate degree was required for admission to the course. After the First World War, forestry became an undergraduate degree subject. Most successful students entered the Indian Forest Service, but there was also a demand for foresters in other colonies across the empire.

By the time he retired, Schlich had raised enough funding to endow a chair of forestry. His successor as head of the School of Forestry was an Indian colonial forester and his former student, Robert Scott Troup. Troup became the founding director of the Imperial Forestry Institute (IFI) in 1924. The IFI's function was to be the central facility for the advanced training of forest officers and a research centre for the 'formation, tending and protection of forests'. Advanced education was provided in the form of one-year bespoke courses, aimed at students from the British colonies, on subjects that ranged across the interests of the academic staff, including systematic botany, silviculture, forest management and wood properties. Troup's *Silvicultural Systems* (1928) is still in print and remains an important forestry textbook. The economic depression of the 1930s saw reductions in both student and staff numbers. Staff remained more numerous than in the Department of Botany, but the separation of the School of Forestry and the IFI proved unworkable. The two were merged in 1939, under the directorship of silviculturalist Harry George Champion, a pupil of Church who taught botany to foresters. During Champion's tenure, the undergraduate forestry course was extended to four years and remained in place for the next three decades.

The reconstruction of the British economy after the Second World War required the British colonies to exploit the natural capital in their tropical forests as never before. Many of the trees in these tropical forests had not been identified and their properties were not known. The IFI, together with the Department of Botany, began the task of cataloguing their diversity, and identifying the timbers with the greatest potential value, particularly poorly known species.

As with the Department of Botany, space for students and research was at a premium in the IFI, which became the Commonwealth Forestry Institute (CFI) in 1947. In the early 1930s Charles Vyner Brooke (the White Rajah of Sarawak) made an opening gift of £25,000 (c.£1.1 million in 2020) to a building fund, which was followed by many other donations from individuals, companies and governments. In 1934 the university allocated a site on South Parks Road for a botany and forestry building. Opposition in the university was apparently fierce: the vice chancellor made the comment that 'some have thought that Foresters should be kept out of Paradise, as Adam and Eve once were, since Forestry is not one of the prime sciences. Others have held that the whole Empire would go to Hell unless a site in Paradise was granted to the Foresters.'

The paradise for botany and forestry, designed by Hubert Worthington, Slade Lecturer in Architecture at the university, finally opened in the early 1950s: two departments surrounding an open quadrangle united by a common lecture theatre. During the 1950s all officers of the Colonial Forest Service were given one year of postgraduate training from the CFI, while similar courses were offered to foresters from former British colonies.

In 1959 the teak expert and former head of the Forestry Commission's Forest Research Station at Alice Holt Lodge, Malcolm Vyvyan Laurie, took over from Champion as head of the CFI. In 1962 the Empire Forestry Conference charged the CFI with undertaking intensive research on fast-growing plantation tree species, especially the Caribbean pine (*Pinus caribaea*). In response, a research group quickly developed that had a wide range of forestry skills, including seed collecting, silviculture methods, genetic improvement and conservation, experimental design and biometrics, and forestry pathology and entomology. Over the next four decades these skills were deployed to address many different practical forestry problems in developed and developing countries.

Nevertheless, by the time the forest mycologist Jack Harley took over the CFI in 1969, many within the university viewed it as moribund – 'an anachronistic relic of colonial days'.[22] Rather than oversee its demise, Harley invigorated it, dramatically improving the science that was done. In 1970 the Departments of Agriculture and Forestry introduced a popular joint undergraduate degree on agricultural and forest sciences. A taught MSc, called 'Forestry and its Relation to Land Use', was introduced in 1972. In the three decades of its existence, the course produced over 460 graduates, ensuring that Oxford-trained foresters occupied most of the senior positions in world forestry well into the twenty-first century. Research strengths in forest pathology (especially

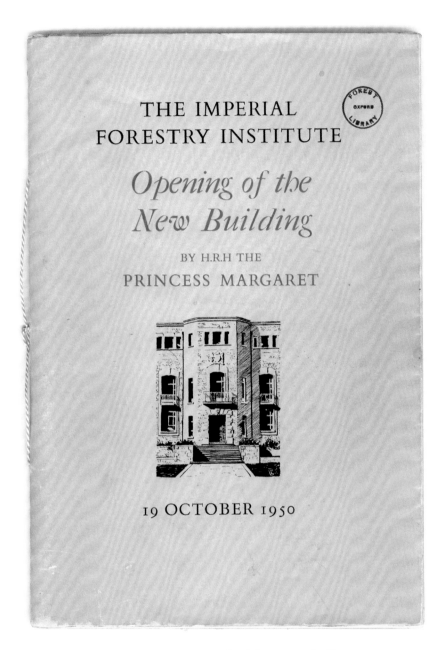

THE IMPERIAL
FORESTRY INSTITUTE

*Opening of the
New Building*

BY H.R.H THE
PRINCESS MARGARET

19 OCTOBER 1950

virology), entomology and tree ecophysiology were built up.

On his retirement in 1980, Harley was briefly replaced by the conservationist Duncan Poore, who was then followed by the forest geneticist Jeffery Burley in 1982. A year later the CFI became the Oxford Forestry Institute (OFI); Burley was the last head of the Forestry Department and the first and last director of the OFI. The latter part of the century saw dramatic changes in forestry at Oxford, driven by

Brochure celebrating the opening of the Imperial Forestry Institute in 1950. Oxford University Herbaria.

both national politics and pressures within the university. Government policies in 1981 forced degrees in botany, zoology and agricultural and forest sciences to merge into a degree in pure and applied biology, which in 1988 became biological sciences. Forest entomology moved to the Department of Zoology in the 1980s, while virology moved out of the OFI to form the Natural Environmental Research Council's Institute of Virology. In 1985 the Departments of Forestry, Agriculture and Botany merged to become the Department of Plant Sciences. During the 1990s national and international interest in the social science aspects of forestry increased, emphasizing the difficulties of sustaining academic forestry in a department driven by the output of plant sciences research. Consequently, as forestry posts became available, they were usually filled by researchers who fitted this agenda. The OFI closed in 2002. In 2011 the Wood Professor of Forest Science was endowed within the department.

Plants in society

From the mid-eighteenth century, people across all levels of British society wanted information about the plants around them.[23] There was a demand for information that went beyond simply how to grow and cook them or how to heal with them. The folkloric gave way to interest in the wonders that natural philosophers were discovering about the lives of plants. Science was appealing to the public imagination. It became something to which anyone could contribute, as writers specifically engaged in the communication of new science emerged in the Victorian period.[24]

From the 1760s, the Linnaean sexual system of classification took hold among naturalists and wider British society, although this interest was driven by people and organizations outside the universities of Oxford and Cambridge. Regional societies of naturalists were established among like-minded people. One of these, the Ashmolean Society, established in 1828, survives today as the Ashmolean Natural History Society of Oxfordshire, following its merger with the Oxfordshire Natural History Society by George Claridge Druce in 1901.[25] A market for information was filled by popular books that translated technical tomes, often written in Latin, into terms that most people could understand. Foremost among these natural history translators were Anglican clergy, a global diaspora of men, many of whom had been trained at Oxford or Cambridge.

Tensions between science and religion were exposed as never before, perhaps most (in)famously at the 1860 meeting of the British Association

Herbarium specimens of spring gentian (***Gentiana verna***) collected by George Claridge Druce at various localities in County Clare, Ireland, in the early nineteenth century. Oxford University Herbaria, Druce s.n.

Murrough
Co. Clare

Feenagh
Co. Clare

Muckinish
Co. Clare

1762

HERBARIUM BRITANNICUM.

GENTIANA VERNA L.

Ballyvaughan Co.Clare

June, 1909   G.C.Druce

G. CLARIDGE DRUCE, M.A., F.L.S.,
YARDLEY LODGE, OXFORD.

for the Advancement of Science, at the recently opened University Museum, where Thomas Henry Huxley and Bishop Samuel Wilberforce discussed the implications of Charles Darwin's recently published evolutionary ideas for human ancestry.[26] The clergyman Charles Kingsley took a more measured approach. After reading *On the Origin of Species* (1859), he wrote to Darwin:

> judging of your book. 1) I have long since, from watching the crossing of domesticated animals & plants, learnt to disbelieve the dogma of the permanence of species. 2). I have gradually learnt to see that it is just as noble a conception of Deity, to believe that he created primal forms capable of self development into all forms needful pro tempore & pro loco, as to believe that He required a fresh act of intervention to supply the lacunas w^h. he himself had made. I question whether the former be not the loftier thought.[27]

Kingsley was a strong supporter of organizations involved in self-education. At Oxford the most prominent autodidact was George Claridge Druce, who always felt stigmatized by both his lack of formal schooling and his parentage.[28] Druce was an enthusiastic, well-regarded public lecturer who took pride in conveying information about plants to audiences as diverse as the landed gentry, university academics, workers' educational associations, natural history societies and schoolchildren.

Charles Daubeny's flirtations with popularization of the Garden included briefly growing the Victoria water lily and publishing *A Popular Guide to the Botanic Garden of Oxford* (1850).[29] In the latter half of the twentieth century, the involvement of the Garden in public education about plants and aspects of plant biology that were of direct interest to the public, such as conservation, increased. Mass appeal and entry charges to the Garden have become essential for its function, even driving its appearance.

In the early twentieth century, the Sibthorpian professor William Somerville was an effective promoter of the role of agricultural research in benefiting British farmers. By the end of the century, Oxford plant scientists were actively engaged in many academic debates of public interest, including acrimonious discussions about the roles of genetically modified organisms in agriculture.[30]

However, the most consistent engagement with getting data from academic research into real-world practice was through the various incarnations of the OFI. Much of the research was published in-house, by-passing traditional publishers, their costs, their conventions and their distribution networks.[31] Typical publications were check-lists of tree species in individual tropical countries, detailed accounts of specific trees, guides to research methodologies and comprehensive reports of research projects. Everything was guided by national and international research agendas, and targeted at specific audiences, whether they were researchers, policymakers, governments, local communities or individual foresters. These publications were augmented by bespoke training courses. Particularly successful long-term courses were developed in research methods and statistics, planning and management, and social and community forestry, with participants coming from more than thirty countries. The core lecturers were OFI researchers, supported by international subject expertise from across the globe. There was also explicit support for the OFI's work from British academic institutions, national and international aid agencies, businesses, the World Bank and the Food and Agriculture Organization.

This approach to publication – where the use of research data and its effective dissemination were of greater value than the accumulation of author credits – was a good collegiate strategy. However, it proved more or less fatal when the wind changed in the late twentieth century, and research was valued only if it was published in peer-reviewed journals that were judged to have a high academic standing.

Oxford botanists were late arrivals at getting their research and ideas into a domain beyond the academic audience, and at effectively engaging with the contributions that plant sciences can make to people's lives. Technological changes have meant that information from research can be made freely available in real time to vast audiences. The challenge for current researchers is to ensure that their message is clear, concise, accurate and relevant to their intended audiences. In addition, the mechanisms for members of the public who are interested in plant sciences to contribute to research agendas through citizen science initiatives have also continued to expand.

Joseph Burtt Davy (1870–1940)[32] was the first curator of the herbarium that was to become the heart of the Imperial Forestry Institute. Burtt Davy was born in Derbyshire and studied agriculture at the University of California in the mid-1890s. For the rest of the century, he specialized in grasses and sedges which he eventually published in Willis Jepson's *A Flora of Western Middle California* (1901).

In 1903 he took up a botanical post, again specializing in grasses, at the Transvaal Department of Agriculture, which later became the South African National Biodiversity Institute. Following the publication of *Maize: Its History, Cultivation, Handling and Uses with Special Reference to South Africa* (1914), Burtt Davy resigned his post. He used the practical knowledge of grasses and plant breeding he had gained to farm commercially in South Africa.

Five years later he was wealthy enough to retire to England and to continue working on *A Manual of the Flowering Plants and Ferns of the Transvaal with Swaziland* (1926, 1932). In 1925 he took up the challenge of a Civil Service post as lecturer in tropical forest botany at the newly formed Imperial Forestry Institute. He worked on *Forest Trees and Timbers of the British Empire*, and trained forestry officers, who were needed by colonies across the British empire. He is commemorated by the name of the eastern and southern African tree genus *Burttdavya*.

Geoffrey Emett Blackman, FRS (1903–1980),[33] an agricultural scientist whose father was professor of plant physiology and pathology at Imperial College, brought robust statistical methods to the solution of botanical problems when he arrived in Oxford.

As a student, Blackman worked at the Rothamsted Experimental Station under the statistician and geneticist Ronald Fisher, studying the effects of fertilizers on plant growth. However, Blackman's father considered him better suited to industry than academia, so he joined Imperial Chemical Industries (ICI), where he worked on the effects of fertilizers and poisons on ecological interactions. By the early 1930s, frustrated by the routine and secrecy of ICI, he took up a lectureship at Imperial College, where he focused his attention on ecological experimentation in natural habitats. During the Second World War he supervised research on wartime agricultural problems.

At the end of the war, Blackman was elected the seventh Sibthorpian professor. He changed the emphasis of agricultural teaching and research in the university from farm management to the science of agriculture. At Oxford he built a successful group focused on the physiology of plant hormones and plant growth. His experiences at ICI made Blackman sceptical of academic involvement with industry, a position that he maintained at Oxford.

7 SEED
Teaching

Modern universities, which originated in the twelfth and thirteenth centuries, spread rapidly across Europe.[1] In the United Kingdom, they emerged from a monastic tradition where like-minded men created communities or guilds to order their lives, defend their status and protect their privileges. What eventually emerged was a new institutionalized group that believed it had the ability and knowledge to either govern or teach others to do so. While universities were initially dominated by clerics, laymen, especially lawyers and doctors, soon began to emerge from them.

In 1713 the English clergyman and poet Abel Evans, who exhorted Bobart the younger to 'Instruct us in thy Mysteries: From Thee, the *Gods* no Knowledge hide', acknowledged the responsibilities of those in Bobart's position to disperse their knowledge: 'Let others, meanly satisfy'd With Partial Knowledge, sooth their Pride.'[2] Evans's list of people who would benefit directly from Bobart's botanical knowledge is likely to have included physicians, students, curious academics, and those who visited the Garden, read his books and talked to, or corresponded with, him.

Peer-reviewed papers in academic journals or in specialist books, the usual routes by which botanical research is published, help disseminate and establish ideas. Research that is unpublished or data that are locked away may as well not have been undertaken or collected. Yet research papers and books are read by very few people. In the short term, a more effective way of spreading botanical ideas may be through how people are taught and trained in the university, and what they subsequently achieve. Despite their uneasy relationship with the university, training in agriculture and forestry were fertile areas for the dispersal of ideas about scientific botany globally (see Chapter 6).

Oxford ragwort (*Senecio squalidus*), depicted by William Baxter (*detail; see p.193*)

THE OXFORD RAGWORT

A mysterious Sicilian plant arrived in Oxford in the early eighteenth century. We do not know how it arrived, but we do know that at least one Italian monk (Francisco Cupani, director of a botanic garden near Palermo), a British diplomat with a botanical bent and an amiable personality (William Sherard), a dowager duchess (Mary Somerset, Duchess of Beaufort) and Jacob Bobart the younger were probably involved with the movement of seeds into England.[3]

From the time of Bobart the younger until the early nineteenth century, Oxford ragwort (*Senecio squalidus*) was a novelty confined to the city's walls and a few parishes looked after by Oxford-trained prelates who wanted a souvenir of their student days. The light, parachute-like ragwort fruits are effective dispersal units. Yet, to be ecologically successful in its new homes, the Oxford ragwort needed time to adapt to its conditions, to find suitable habitats in which it could grow and to break out of Oxford.

The 1840s brought Oxford the railway, which, as Druce noted, with its track ballast 'furnished the plant with a replica of the lava-soils of its native home in Sicily'. He went on to observe that 'the vortex of air following the express train carries the fruits in its wake. I have seen them enter a railway-carriage window near Oxford and remain suspended in the air in the compartment until they found an exit at Tilehurst [c. thirty-five kilometres from Oxford].'[4]

In this way the ragwort escaped Oxford, and spread via the railways through western Great Britain. The destruction of the Second World War produced more habitats for Oxford ragwort, and a second colonization front opened in the east. Today, Oxford ragwort's distinctive yellow, star-shaped flowerheads are familiar across the country, wherever well-drained, human-made habitats abound.

Carl Linnaeus gave the species its Latin name, based on samples collected from the walls of the Botanic Garden, which were apparently given to him by Johann Dillenius in the 1730s. However, the first complete account of the plants of Oxfordshire, published in 1794 by John Sibthorp, failed to link the ragwort on the city walls with that described by Linnaeus.

Oxford ragwort is more than a plant that happened to spread after leaping over a garden wall; it is a model for studying plant evolution. Crosses with our native groundsel (*Senecio vulgaris*) produce sterile hybrids, but sometimes a few plants, following complex genetic changes, are fertile. In the last 150 years, three of these hybrids have been formally described as new species. The Oxford ragwort is itself a hybrid between two Italian species: one parent is endemic to the top of Mount Etna; the other is widespread in southern Italy. The adaptation of Oxford ragwort to life continues. For example, Oxford ragwort arrived in Scotland as recently as fifty years ago, yet populations show genetic differences associated with survival at low temperatures when compared with those from southern England.[5]

SENECIO SQUALIDUS. INELEGANT RAGWORT. ☉

Oxford ragwort (*Senecio squalidus*), depicted by William Baxter in his *British Phaenogamous Botany* (1834). Bodleian Library, Sherardian Library of Plant Taxonomy, (4.1) BA2Ee, t.52.

Today, plant scientists working in Oxford spread information to audiences across the world using a broad range of media. Their message is clear: plants are important in our lives, and worthy of sustained curiosity from good researchers and teachers. However, Oxford's success at both the generation of botanical knowledge and its dispersal over four centuries has been very patchy.

Early botanical teaching

By the early seventeenth century, a university education at Oxford offered three formal courses: theology, law and medicine. Men were prepared for the cloth, bench or consulting room. For those interested in natural history who wanted or needed the benefits of a university education, opportunities were restricted primarily to medicine.

At Oxford, as elsewhere, teaching about and investigation of plants was confined by its university context, the classical origins of botanical knowledge and its association with medicine. The strong separation of gown and town may also have potentially reduced the 'value' of knowledge from non-academics. Bobart the elder was not a member of the university, so if he was the author of the 1648 Garden catalogue, his anonymity may have ensured that the book would be acceptable within Oxford. 'Pure' botanical knowledge, acquired by thought and from books, was prized above 'applied' knowledge, acquired through practical experience.

Moreover, the medical origins of university-based botanical training led to further tensions between pure and applied science, with the observation and cataloguing of plants prized above experimentation.[6] In contrast, zoology, which also originated in medicine, was concerned with the ordering of animals along the *scala natura*, an ancient Greek idea that located humans at the top of the zoological ladder.[7] That is, zoologists were particularly interested in the similarity of humans to other animals. Consequently, when zoological education emerged from medicine in the nineteenth century, experimental and observational studies were given similar status.

Early modern botanic gardens, as places where medicinal plants were labelled with their correct name, had the potential to become merely narrow, living catalogues of useful plants. However, the mid-seventeenth-century Botanic Garden created by the Bobarts was more interesting than that: rather than being a collection of plants focused on medical training, it came close to John Evelyn's concept of a 'Philosophico-Medical Garden'.[8]

Morison and the Bobarts as teachers

The first formal lecture course in the Botanic Garden began on Monday, 5 September 1670, with the newly inaugurated Regius Professor of Botany, Robert Morison, standing behind a table of plants at the centre of the walled Garden.[9] His audience may have expected a lecture on medicinal plants, as Morison's inaugural lecture had taken place a few days earlier, on Friday, in the school of medicine.

The new professor had the Bobarts to thank for the selection of choice plants from across the known world with which he stocked his table for demonstration purposes. September was usually a poor time of the year to begin a course of botanical lectures in an English garden, but for Morison the abundance of fruits rather than flowers was ideal. During his exile in France, he had conceived and started to develop a fruit-based classification system for all plants. Morison had little interest in laboriously cataloguing the medicinal uses of plants. Consequently, his audience was probably introduced to his ideas about plant classification, his reflections on the herbal tradition and the limitations of earlier authors, especially the early seventeenth-century work of the Bauhin brothers.

Morison's opening lecture was authoritative and well attended, and was perhaps boosted by an audience curious as to what the first Regius Professor of Botany in the country had to say standing in a Garden that had only started to be planted three decades earlier.[10] Doubtless, the Bobarts were present, dancing attendance on their new professor.

We know nothing of how Morison's audience reacted to his lecture. What little we do know of his manner tells us that Morison was an engaging teacher, although his strong Scottish accent marred his eloquence for the Oxford antiquarian Anthony Wood: 'though a master in speaking and writing the Latin tongue, yet [he] hath no command of the English, as being much spoyled by his Scottish tone.'[11] Wood reported that when James Stuart (later James II of England and Ireland and James VII of Scotland) visited the Garden in 1683, Morison's mangled English provoked laughter. A seventeenth-century diarist and former Oxford student, the Reverend John Ward, reported that Jacob Bobart the elder said of Morison that 'ye whole world yields not ye like man, hee never heard a man talk att yt. gallant rate in his life'.[12]

Morison's course was planned to occur thrice weekly for two sessions, each five weeks long. The first session started in September, and the second session the following May, when no doubt flowers were the focus. In September 1671 Morison started to repeat his course, but by the following May his attention had been diverted away from teaching into the research necessary for his *Historia*.

The tension between professorial time for teaching (in lectures and tutorials) and professorial time for research emerged early in the botanical history of the university. The professor's teaching activity became sporadic; most of it was probably handed over, at least informally, to Jacob Bobart the younger. The delegation or abandonment of professorial teaching responsibilities became a familiar feature of the Garden for the next 250 years. When Morison died suddenly in 1680, Bobart the younger became the de facto professor, although without the formal recognition.

Bobart the younger's interest in and dedication to teaching botany may have been acquired from seeing his father giving instruction in the Garden. In the early 1660s Ward records how he learned about the plants growing in the Garden, the locations of unusual plants growing about Oxford and perhaps even the art of pressing plants and making a 'Botanologicall Book' from Bobart the elder.[13] The latter's teaching, given his adherence to Morison's classification system, is likely to have followed a similar format to that of the Regius professor. The son, modest about his knowledge, also appears to have made a great impression on the students he taught through peripatetic lessons within the confines of the Garden.[14] His most illustrious pupil, the law student at St John's College, William Sherard, went on to devote most of his life and fortune to botanical research. Sherard also remained a friend and defender of Bobart to the end of his days. When he died in 1728 Sherard left his library, his herbarium and the bulk of his fortune to establish the Sherardian Chair of Botany. He is also likely to have been an important node in social networks that enabled Bobart the younger to engage with elite members of late seventeenth-century society.

With the death of Bobart the younger in 1719, a gap opened up in both the botanical teaching and the reputation of the university. In 1719 the university appointed Edwin Sandys as professor of botany, and in 1724 Gilbert Trowe; neither man left a botanical legacy. In 1723 the naturalist Richard Bradley had lobbied the future president of the Royal Society

Early eighteenth-century oil portrait, by an unknown artist, of William Sherard, who established the Sherardian Chair of Botany at Oxford University. University of Oxford, Department of Plant Sciences.

Hans Sloane to promote his candidature for the Oxford post so that botany 'will be put on a new foot and the physick garden brought into order'.[15] Bradley's interests in the application of experimental botany to horticulture no doubt included teaching, but his bid was unsuccessful. In 1724 Bradley was made the first professor of botany at Cambridge University, where he was considered duplicitous and a lazy teacher by his two immediate, long-serving and traditionally minded successors. Less partisan analyses show that Bradley contributed to developments that ultimately led to modern ecology and an understanding of the causes of infectious disease.[16]

Setting the rules for teaching

In 1736 the foundation of the Sherardian chair gave the university, through the committee that had oversight of the Garden, the opportunity to formalize the teaching duties of the professor. He was to

> begin his Lectures about ye middle of March ... once a Week ... till ye End of Aprill ... & twice in ye Week during ye Months May, June, July & August, unless he shall think fit to absent himself on his own or Garden-Affairs ... & then in September to resume his Lecture & read only once a Week till ye Season is entirely over. The Length of ye Lecture to be calculated in Proportion to ye Number of Plants growing in the Garden ... The Days & Hours of demonstrating to be such as the Professor himself shall judge proper.[17]

The professor was to 'annually open his Lectures in ye Spring with a short Speech in Latin ... close them in ye same Manner in Autumn & also make a Botanick Harangue once every Year in ye Physick School'.

Notwithstanding the formalization of teaching duties, the eighteenth century was the age of the absentee professor, whether physically or intellectually, at both Oxford and Cambridge.[18] Arguments of bias against the emerging sciences in institutions dominated by theologians may have some merit. Sherard specifically stated the electoral board for the Sherardian professor must have a member of the Royal College of Physicians; he feared the university would fill the post with a theologian after Dillenius.[19] However, that is not the whole story. At Oxford, science was apparently tolerated if it did not threaten the primary purpose of

The timing and length of botanical lectures, which were part of the teaching duties of the Sherardian professor, were first laid out by the committee responsible for the Botanic Garden on 7 February 1736. Bodleian Library, Sherardian Library of Plant Taxonomy, MS. Sherard 1, f.5r.

Mem.^{dum}   February 7th 1735/6

    In a Meeting of a Committee appointed
for the Care of the Physick Garden it was
this Day agreed.

    1st That ye Professor of Botany shall annu-
ally begin his Lectures about ye middle of
March or sooner if a forward Spring to be
continued once a Week from that Time till
ye End of Aprill, or oftner if the Number of
Plants flourishing at that Time shall require,
& twice in ye Week during ye Months May,
June, July & August, unless he shall think fit
to absent himself on his own or Garden-Affairs,
which by ye Decree he is allow'd to do for six
Weeks some Time in ye Summer, at any Time
of ye aforesaid Months, & then in September
to resume his Lecture & read only once a
Week till ye Season is entirely over. The Length
of ye Lectures to be calculated in Proportion
to ye Number of Plants growing in the Garden,
so as that ye whole Garden may be demonstrated every Year. The Professor not to be absent
above one Week at a Time during ye Months
of May, June, July, & August, without the special leave of the Committee. The Days
& Hours of demonstrating to be such as the
Professor himself shall judge proper.

the university – the defence and promotion of the Anglican Church.[20]

There is no trace of whether the first Sherardian professor, Johann Dillenius, contributed to teaching. His pupil, Humphrey Sibthorp, who held the chair for thirty-seven years, is notorious for having given only one – unsuccessful – undergraduate lecture, although our only evidence comes from the founder of the Linnean Society of London, James Edward Smith, who disliked Sibthorp.[21] In 1760, a seventeen-year-old Joseph Banks, the future intellectual architect of the Royal Botanic Gardens, Kew, arrived at Oxford, and soon became bored with the traditional subjects on offer.[22] Botany interested him, but Sibthorp did not teach; instead he approached his counterpart at Cambridge, John Martyn, to recommend a tutor for Banks.[23] In July 1764 Martyn's choice, Israel Lyons, delivered a successful course of botany lectures to sixty students at Oxford, at Banks's expense. Lyons was probably the first person at Oxford to teach botany using the Linnaean system.[24]

Botanical teaching at Cambridge in the eighteenth century was little better than at Oxford; both institutions appeared dormant, the chairs merely sinecures. John Martyn,[25] a wealthy businessman and talented botanist, who had formed a London-based botanical society with Dillenius in 1721, was elected to the Cambridge botany chair in 1733. Two years later, Martyn had lost interest in Cambridge and became an absentee professor, returning to his business interests in the capital. He finally relinquished the post in 1762, in favour of his son. For at least the first half of his six-decade-long tenancy of the chair, Thomas Martyn, 'a fortunate man to whom everything [had] been easy, but whose talents were mediocre',[26] was nevertheless a dedicated teacher of Linnaean botany who inspired students in both the field and lecture room.[27]

The parallels between Oxford and Cambridge botany in the eighteenth century are striking: father–son relationships in professorial successions; protracted joint occupations of chairs (the Sibthorps for forty-nine years, the Martyns for ninety-two years); academic lethargy; and the enthusiasm of the sons for Linnaean botany.

Teaching Linnaean botany

In 1783, once John Sibthorp had convinced his father to resign the Sherardian chair in his favour, it appeared that the son would follow in the father's footsteps. John left Oxford in 1784 with a generous travel grant won for him from the university by his father, to embark on an

John Sibthorp gave a thirty-lecture course on Linnaean botany to students between c.1788 and c.1793, in which he made regular comments about the poor conditions for growing glasshouse plants. Bodleian Library, Sherardian Library of Plant Taxonomy, MS. Sherard 219, f.67r.

Besides these Libraries which are as it were the Repositories of the Labors of the Botanists; our Inquiries are excited by the numerous Botanic Gardens with which this Country is enriched, both public & private. At the Head of these stands the Royal Botanic Garden of Kew. The Catalogue of this Garden lately published under the Title of Hortus Kewensis, ascertains its Claims. To supply this Garden Botanists are sent out at a great Expence to examine & procure the Rarities of the most distant Countries. The greater Number of these Plants require an artificial Shelter — the extent of which demands a royal Munificence to support & maintain. Academic Gardens, tho' greatly inferior in Magnificence & Splendour, to those supported by Royal Expenditure, may be considered as the more Useful Schools of Botany — not under the Restrictions of Royal or private Collections — they are at all times open to the Student, & their Object is to inform as well as amuse. Picturesque Beauty is not merely considered, but Method & Order as far as they conduct to a systematic Arrangement, must be preserved. The Method we have chosen is that of Linnæus which we are persuaded is the best — we have reserved this Method — admitting only some few alterations in respect to the distribution of some of the Genera (which Linnæus himself probably would have made) The Names are entirely the Linnæan except of some new species which have been either discovered or distinguished since his time. These Names for the Benefit of the Botanical Students are in plainly legible Characters on Labels affixed to the Plants

extensive European tour.[28] He did not return to Oxford until 1787, and left for the last time in 1794. But for the seven years Sibthorp was at Oxford, he gave regular undergraduate lectures.

Sibthorp's thirty-lecture botany course,[29] delivered to an audience of young men who were interested primarily in medicine and agriculture, emphasized the benefits of studying botany. Besides the economic and medical value of plants and the discovery of new information, Sibthorp argued, the botanist 'will re-establish his Health deranged by the Confinement of the Closet'.[30] The extant lecture notes are a rich source of information about the botanical knowledge Sibthorp thought would be useful for students in the late eighteenth century.

During his study of medicine in Edinburgh, Sibthorp became fascinated with two elements of Linnaean botany – binomial names and the revolutionary sexual classification system – through the teachings of John Hope.[31] Sibthorp became an enthusiastic teacher of the Linnaean system. At the opening of his third lecture he proclaims: 'We have now arrived at the most interesting Period of the Progress of Botany, when the bold but systematic Genius of Linnaeus forged as it were a Chain, which encompassed the whole of Nature.'[32]

Sibthorp appears to have started his lecture course soon after his return to Oxford from his pioneering botanical explorations of the eastern Mediterranean in 1787. When he started to teach these ideas, they were effectively new in the university. Linnaeus had been dead for about a decade, while some of the ideas had been in circulation for near fifty years and had been enthusiastically taught by Thomas Martyn in Cambridge.[33] Oxford was emerging from botanical hibernation.

Sibthorp's anecdote-laced lectures, which took place in the Botanic Garden, focused on the uses of plants, especially their role in agriculture and as foods and medicines, which is hardly surprising given that his audience were likely to have been landowners or prospective physicians. The order of the lectures will be familiar to any reader of botanical texts. They opened with three lectures on the history of botany from the 'earliest People' through Linnaeus to Sibthorp's European contemporaries, before students were plunged into three lectures on the details of plant structure. The remaining lectures were devoted to the systematic consideration of the twenty-four divisions of Linnaeus' sexual classification system, although the examples chosen to illustrate the classes were those most likely to hold students' attention.

Sibthorp coloured his lectures with tales from his travels, confirming or refuting the accounts of other travellers with whom his students were probably familiar. For example, when discussing the Diandria and *Jasminum officinale*, Sibthorp's students were told

> the Turks are particularly fond of this Tree & greatly esteem the Jessamy Stock for the Tubes of their pipes when these are long, straight, & of a good Colour. They are sold at the Enormous Price of a 100. piastres per Stock – about. 10. £ of our Money [*c*.£860 in 2020].[34]

He used all the material at his disposal to illustrate his lectures, including books from his personal library, specimens from the university's herbarium, for which he was responsible, living plants from the Botanic Garden and watercolours completed by the botanical artist Ferdinand Bauer, who was working in Oxford under his direction. Students were shown Bauer's watercolour of the olive (*Olea europaea*) to illustrate the tree's fruits, along with the reproof: 'Our Olive Tree tho' it produces its flowers, seldom ripens its Fruits & that you may have a more perfect Idea of it I shall show You the fructification in a drawing.'[35] This watercolour was later reproduced in the *Flora Graeca*, and its use as a teaching prop may explain why the original is now rather grubbier than many other watercolours in the collection.

Sibthorp emphasized the scientific importance of the Botanic Garden in Oxford compared to the fledgeling Kew, and made passing remarks about the university authorities, with whom he was battling for funding:

> Academic Gardens tho' greatly inferior in Magnificence & Splendour to those supported by Royal Expenditure may be considered as the more useful Schools of Botany. – not under the Restrictions, of royal or private Collections, they are at all Times open to the Public, & their Object is to inform as well as amuse. Picturesque Beauty is not merely studied, but Method & Order as far as they conduct to a Systematick Arangement must be preserved.[36]

However, he was well aware of the limitations of the Garden for teaching purposes: 'in the first Order Monogynia we find the Caper Bush Capparis

– but this plant unfortunately I cannot demonstrate as we have it not at present in our Garden'.[37]

Sibthorp was enthusiastic about the changes happening in the planting arrangement of the Garden, which was coming closer to his ideal for teaching purposes:

> Since we met last we have continued our Arrangement – & we have chosen that Method which we are persuaded from Experience & Conviction is the best. The Quarter now before us contains all the British perennial Plants, whose Situation does not require a particular Position. I mean the Alpine Plants, & such as grew in very moist Situations – these we have contrived under the Cover of a Wall facing the North to place in Such a Situation as they naturally grew in – that we might as much as possible observe their Natural Growth, neither disguised nor distorted by Art.[38]

Despite his primary interest in flowering plants, he was keen on mosses and ferns, which 'fortunately for the Botanist ... flower at the Season of the year, when there are few other plants to engage his Attention'.[39] Sibthorp was particularly fascinated by the experimental work of the 'ingenious' German botanist Johann Hedwig on the sexual reproduction of mosses, and the comparison of Hedwig's ideas about bryophytes to those of his early eighteenth-century predecessor Johann Dillenius:

> The Obscurity attending the Fructification of the Ferns as well as some other Genera of this Class has lately been in great Measure dispelled by the deep sighted Researches & the indefatigable Industry of the ingenious Hedwig. His physiological Discoveries relative to the Structure of these plants rank among the most interesting Discoveries of this Century.[40]

Callistephus chinensis, **from Jacob Dillenius' personal hand-coloured copy of his** *Hortus Elthamensis* (1732). Many of these plants, from glasshouses in James Sherard's garden at Eltham, came to the Botanic Garden and were used by Sibthorp in his lectures. Bodleian Library, Sherardian Library of Plant Taxonomy, Sherard 643, t.34.

As a conscientious teacher, Sibthorp closed his lecture series by stating:

> we are now arrived at the last link of the Vegetable Chain – & with this terminate our Lectures but tho' these are finished – my office of a Professor still continues – & the Botanical

Aster Chenopodii folio, annuus, flore ingenti speciofo. Elepl China delatus.

Student will find me no less ready to assist his Enquiries in the Time of Vacation than in the Hour of Lecture.[41]

Following Sibthorp's sudden death in 1796, botanical teaching once more dipped into torpor. The new Sherardian professor, George Williams, reworked Sibthorp's lectures, but little new was added, despite the pace of botanical change happening across Europe. In 1813 the newly appointed Garden superintendent William Baxter eagerly assumed responsibility for the botanical teaching, which he continued until after the arrival of Charles Daubeny, as the fifth Sherardian professor, in 1834.[42]

Finding space to teach botany

When Charles Daubeny became Sherardian professor he already held the chemistry chair.[43] He was well known and respected as a teacher in the university, and he understood how the collegiate system worked. He was also aware of the inadequate facilities available for natural sciences research and teaching, unlike some of his successors.

Daubeny considered science as complementary to literary and classical studies, and early in his career became concerned that teaching in the natural sciences at Oxford was falling behind other centres in the United Kingdom and Europe. As early as 1822, against prevailing opinion, he argued that the natural sciences should be part of the elementary liberal education for all undergraduate students. By the mid-nineteenth century, an education system training young men to face the challenges of running an empire appeared to approach Daubeny's views. A School of Natural Sciences was established in 1850.[44]

The ways in which scientific research was carried out, and how science was taught, were changing nationally and internationally. Men started to be paid as professional scientists, although the stigma of professionalization did not disappear till the end of the Victorian era. The university was also concerned that Oxford teaching should not become vocational, or be associated with professionalization, except through the traditional disciplines of medicine and law.

In the mid-1820s Daubeny proposed that the university accommodate all professors of natural sciences at a single site which, with the creation of the School of Natural Sciences, eventually led to the establishment of the University Museum on Parks Road between 1855 and 1860.[45] Most natural science disciplines were therefore accommodated, away from the traditional centre of the university, in a building where the primary focus was teaching. However, even with the additional space, there was still a dire shortage of the laboratory space that was necessary for the natural sciences to flourish.

The University Museum, whose remit was to present 'knowledge of the great material design' to everyone, acknowledged botany with the floral motifs on its column capitals, drawn from models in the Botanic

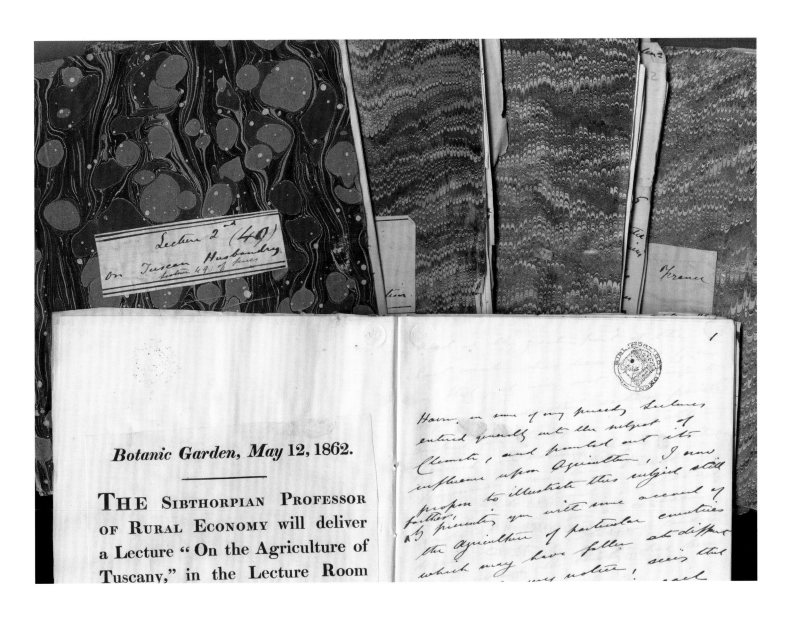

Garden.[46] However, botany had its own cathedral at the Garden, located opposite Magdalen College. The distance between Garden and museum – just over one kilometre – proved a stark divide.

In 1853 Daubeny highlighted the improvements he had made and was continuing to make to the estate of the Botanic Garden, largely at his own expense, so that botany could 'be studied in this place without requiring any further augmentation to our existing means of instruction'.[47] Such optimism was misplaced.

Fewer than twenty years later, the university was faced with a choice. The Science Area around the University Parks had prospered but, without Daubeny's energy, the study of plants was once again floundering. A proposal was made that botany join its intellectual allies in the museum, and that part of the University Parks be used to create a new, freehold botanic garden.[48] Joseph Hooker, director of Kew, was consulted but he thought the Garden would cost more to move than to improve. The sixth Sherardian professor, Marmaduke Lawson, eventually concurred, dividing his arguments to the University Council against such a move into the 'Real' and the 'Sentimental'. Among his 'Real' arguments were the good growing conditions for plants and low human population levels at the Garden – he was concerned with the multiplication of residential properties in North Oxford. Sentimental arguments, to which he attached little value, included the Garden's age, architecture and setting.[49] The university accepted the arguments and agreed to fund changes to the Garden site. A new classroom and laboratory were built to the west of the Danby Gate, and the herbarium was modified to a lecture room. A home was created for the Department of Botany. However, the architect of the unification of the natural sciences, Henry Acland, thought the decision misplaced, responding tartly that botany was 'on a leasehold, apart for the rest of the Scientific apparatus of the university'.[50]

Teaching experimental botany

Isaac Bayley Balfour, the seventh Sherardian professor, resigned in 1888 to become professor of botany at Edinburgh and Regius Keeper of the Royal Botanic Garden, Edinburgh. He handed over to his successor, Sydney Vines, a library, herbarium and laboratory. The German botanist Selmer Schönland, former curator of the Fielding herbarium, detailed, with a certain degree of pride, the Botanical Laboratory and its facilities.[51]

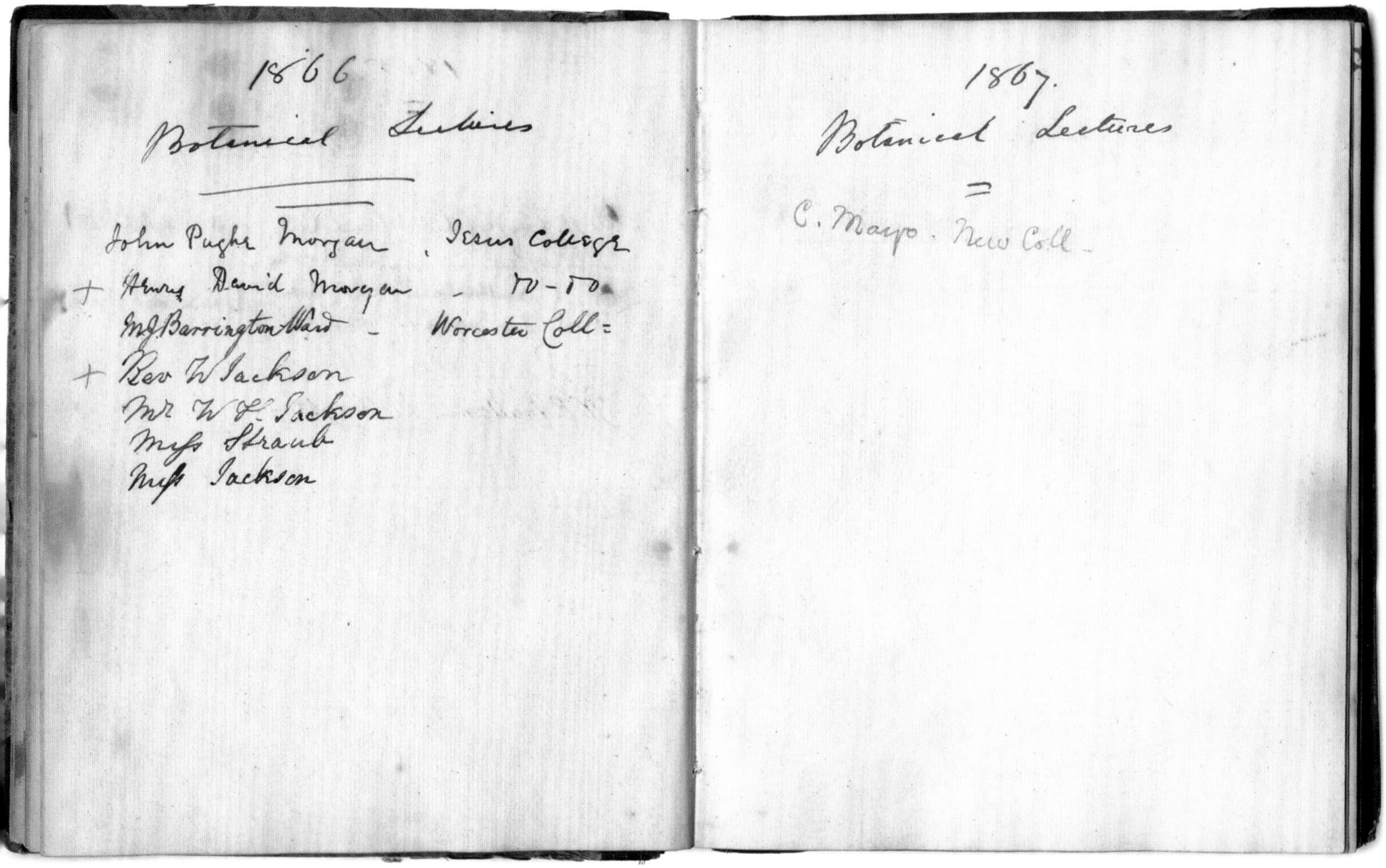

1866

Botanical Lectures

John Pugh Morgan , Jesus College
+ Henry David Morgan — Do — Do.
M.J. Barrington Ward — Worcester Coll=
+ Rev W Jackson
Mr W H. Jackson
Miss Straub
Miss Jackson

1867.

Botanical Lectures
=
C. Mayo. New Coll —

The physiological laboratory, lined with shelves of glassware and reagents, had a 'big cultivator, in which plants can be grown at constant temperatures' and double-wall jars 'which can be filled with colored fluids so as to grow plants with colored light'. The professor's room was used to store delicate instruments, such as 'an auxanometer, a galvanometer, a klinostat, chemical balance, microscopes, microtomes, polariscope, microspectroscope, a magic-lantern'. A darkroom was equipped with a 'very good microphotographic apparatus by Zeiss'. In the lecture room, there was a lecture table and benches for the students to sit at, together with cases for a 'large collection of systematically arranged diagrams and drawings for use in lectures'. Also in the room

Attendance register of an 1866 botanical lecture course in the university which records the names of two of the first women formally taught botany at Oxford University. Oxford University Herbaria.

was an 'ingeniously constructed' piece of apparatus for 'growing algae and other organisms in sea water and fresh water', which had 'not yet worked very well'.[52] Schönland emigrated to South Africa where he founded the Botany Department at Rhodes University and was a leading light in the Botanical Survey of South Africa.[53]

The morphological laboratory, more spacious than its physiological counterpart, was the primary practical teaching space for 'preliminary men' (elementary students). Floor-to-ceiling, south-facing windows had 'quite plain and strongly built', gas-lit tables in front of them to accommodate twenty students.[54] At the room's centre was a general-purpose bench with a preparation area along the back wall. Walls were lined with cupboards containing the usual impedimenta of practical teaching, together with a 'pretty large collection of systematically arranged materials for investigation, chiefly preserved in spirit'.

In an upper room above the lecture room that was known as the Sherard Room was the museum, established at the Garden by Daubeny and Baxter in 1859, with its 'large collection of models and specimens (both in spirit and dry)'. This room was a 'place where the material necessary for showing in lectures are kept, not as a place for the instruction of the public'. The museum, which was accessed by an awkward spiral staircase, had fallen into disuse on Daubeny's death. Instruction of the public took place either at the University Museum or when the Garden was open on Sunday afternoons.[55]

By the mid-1850s botanical teaching in the United Kingdom had become little more than the study of classification focused on flowers. The approaches adopted by Vines and colleagues from other British institutions were a revelation. These men wanted a new botany, where students were given the tools and knowledge to study any plant, from algae and fungi, through mosses and ferns, to conifers and flowering plants. Their teaching did not focus on classification but encompassed comparative anatomy, morphology and physiology, and students were encouraged to look and discover for themselves rather than accept the assertions of their teachers. Vines's translation of several key German textbooks also introduced English readers to Continental, especially German, botanical methods.

Vines arrived at Oxford having transformed botanical teaching at the University of Cambridge, where he was noted for his enthusiasm, dedication and care. One appraisal of his lecturing style was 'easy and effective, with occasional flashes of humour'.[56] His *Lectures on the Physiology*

Teaching poster of the hair moss *Polytrichum* from Arnold and Carolina Dodel-Port's *Atlas der Botanik* (1878–83). Oxford University Herbaria.

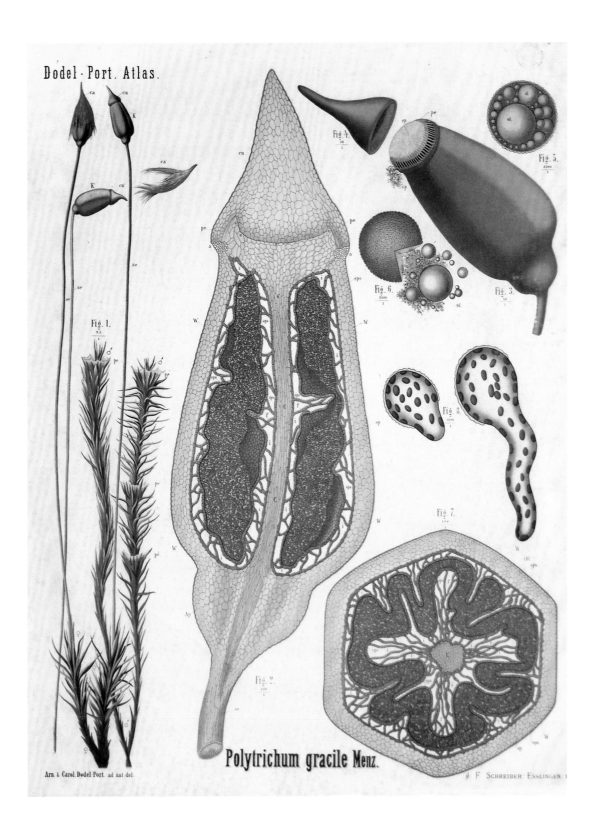

Polytrichum gracile Menz.

F. Schreiber, Esslingen.

of Plants (1886) showed that there was an appetite for the new botany, which led Cambridge to provide him with a large enough space for classes of up to a hundred students. Vines found Oxford students using his co-written textbook *A Course of Practical Instruction in Botany* (1885–7) in practical classes. His co-author, Frederick Bower, one of his students at Cambridge, went on to become professor of botany at Glasgow University. The book was 'produced intentionally without the support of detailed illustrations; for these are apt to blunt the keenness of search, by substituting the observations of others for the personal experience of finding out for oneself'.[57]

In the 'most compact, beautiful, and historically the most venerable'[58] botany department in the country, however, Vines lost his fire, drifting into 'placid efficiency'. It is claimed that, as a teacher at Oxford, he 'put more undergraduates off botany and careers in botany than any other Sherardian professor in history', and he refused to accept women students other than those assured of first-class degrees.[59] Arthur Harry Church, who took an Oxford degree and became Vines's demonstrator in 1894, thought himself fortunate to have learned his botany before Vines came to Oxford.[60] Over the next fifteen years, as Vines and Church taught botany to elementary and advanced students, their relationship soured; Church grew to despise Vines and the type of botany for which he stood.[61]

In 1896 lobbying of the university by the natural science departments in the University Museum for additional space provoked Vines to break his silence about the state of accommodation for the Department of Botany: 'the Botanical Department of this University is more inadequately provided with accommodation in the way of suitable buildings than in any other institution, where Botany is recognized as an academic subject, in the United Kingdom.'[62] As a minimum, he demanded that an additional storey be added to the existing accommodation, whose cost he estimated to be 'not over £1,000 [c.£66,000 in 2020]' – although, even with this addition, his department would be less well served 'compared with other subjects, such as Chemistry, Zoology, &c.'. Apparently even Vines 'never had a decent private room, nor even a work room, let alone a "reception room" like any decent professor'.[63] Furthermore, the financial arrangements that had evolved between the Garden and the Department of Botany were Micawberesque.[64] By the first decade of the twentieth century, student numbers were growing, augmented by forestry students, making conditions at the Garden even more intolerable.

Papier-mâché, paper and wood teaching models of autumn crocus (*Colchicum autumnale*) and cypress spurge (*Euphorbia cyparissias*), made by Robert and Reinhold Brendel in the late nineteenth century. Oxford University Herbaria.

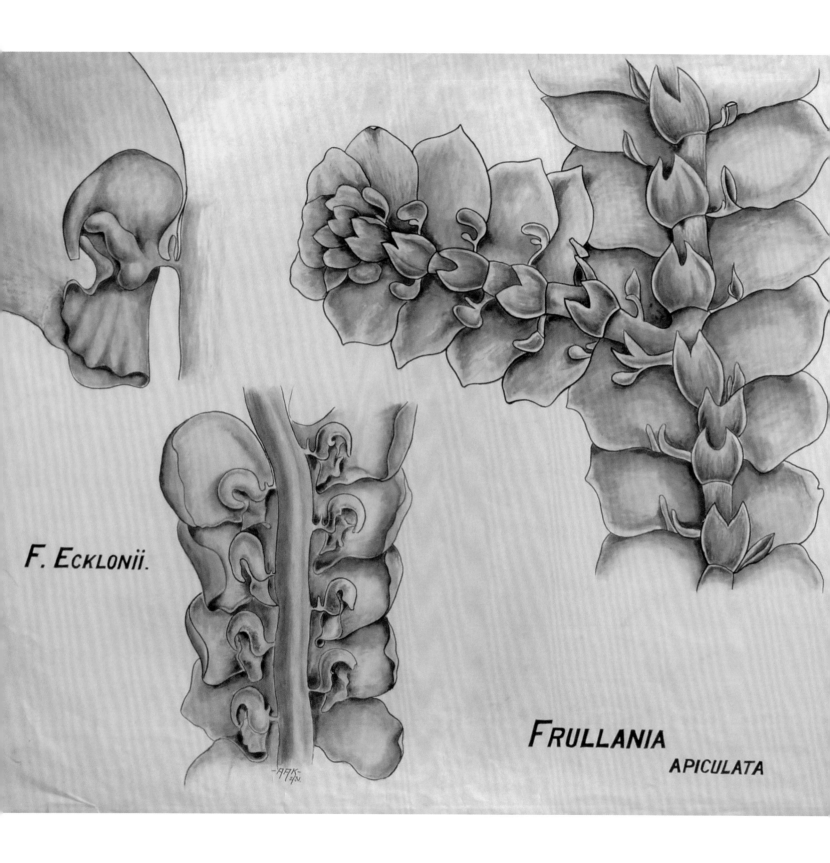

F. ECKLONÏÏ.

FRULLANIA

APICULATA

Botanical teaching between the wars

Frederick William Keeble took up the position of ninth Sherardian professor in 1920. His interests in horticulture were at odds with the direction in which academic botany was progressing in other institutions, especially at University College, London, and at Cambridge; for example, he rejected mathematical and experimental approaches.[65] Although Keeble won additional laboratory space from Magdalen College by taking over the laboratory Daubeny had built, the Department of Botany was barely functional when Arthur Tansley arrived in 1927. Most botanical teaching responsibilities – lectures, practical classes and tutorials – became the de facto responsibility of one man, Arthur Harry Church.[66]

Tansley brought in new, talented academic staff with progressive ideas, and increased the numbers of students. The range of subjects being taught broadened from systematics and physiology to include genetics, mycology and ecology. However, his hopes for more space for teaching and research were twice dashed.[67] In 1930 a grant application to a trust external to the university for creating and equipping a new building and an experimental garden in the Science Area on South Parks Road was rejected. When George Claridge Druce died in 1932, he bequeathed approximately £91,000 (c.£4.7 million in 2020) – almost the amount Tansley had requested two years earlier – to the university. For the university and for Tansley, the bequest came with unwelcome conditions – Druce wanted the money to be used for a Systematics Institute and to care for his herbarium.[68] Four years of legal wrangling ensued before the university finally accepted the money but not all the conditions. Tansley did not benefit from it,

above **Botany Department lecture theatre at the Botanic Garden in the 1930s**, with teaching diagrams hanging from the walls and botanical models crowded in cupboards. Oxford University Herbaria.

opposite **Teaching poster showing details of the structure of liverwort**, drawn by a technician (AAK) in the Department of Botany in 1931. Oxford University Herbaria.

1932 MICHAELMAS TERM

Subject.	Lecturer.	Time.	Place.	Course begins.
§BOTANY.				
Elements of Plant Biology. (Subjects for the Preliminary Examinations in Natural Science and Forestry.) (Fee, £4 with Practical Work.)	Sherardian Professor of Botany, A. G. TANSLEY, M.A.	M. T. W. 10	Botanical Laboratory.	M 10 Oct
Elements of Plant Biology (Practical Work) ..	W. H. WILKINS, M.A.	M. T. W. 11–1	,,	
Introductory Class for Forestry Students. (Fee, £1, including Ecology and Field Classes.)	H. BAKER, M.A.	Th. 10–12	,,	Th. 13 Oct
Elementary Plant Ecology. (For Forestry and first year Honour Students.)	The PROFESSOR	Th. 12, F. 10	,,	Th. 13 Oct
Field Classes. (For Forestry and first year Honour Students.)	The PROFESSOR, H. BAKER, M.A., and A. R. CLAPHAM, M.A.	Th.2, F.11 or 2 (First half of Term only)	,,	
Elementary Anatomy and Physiology (First year Honour Students) (with Practical Work).	The PROFESSOR and A. R. CLAPHAM, M.A.	M. S. 9	,,	/T.
Algae (for first year Honour Students) (with Practical Work).	A. R. CLAPHAM, M.A.	W. 10–1	,,	
Cytology (with Practical Work) ..	A. R. CLAPHAM, M.A.	M. 9.15	,,	
Genetics ..	R. SNOW, M.A.	M. 12	,,	M. 7 Nov.
Gymnosperms (with Practical Work) ..	A. R. CLAPHAM, M.A.	T. 9.15	,,	
Distribution of Plants and Vegetation ..	The PROFESSOR ..	W. 11.45	,,	
Mycology (with Practical Work) ..	W. H. WILKINS, M.A.	Th. 10 (all day)	,,	
Field Class (Mycology) ..	,,	W. 2	,,	
Physiology (with Practical Work) ..	W. O. JAMES, D.Phil.	F. 10 (all day), S. 10–1	,,	F. 14 Oct

§ Names received on Saturday, 8 October, between 9.30 and 1, when the Sherardian Professor will be glad to advise students on their future work. The Sherardian Professor particularly requests all students entering the Department to see him during these hours, or, if this is impossible, as soon as possible during the first week of Full Term.
A fee of £6 covers all lectures and courses of practical work given in the Department.

1933 HILARY TERM

Subject.	Lecturer.	Time.	Place.	Course begins
§BOTANY.				
General Course. (Subjects for Preliminary Examination in Forestry and for first year Honour Students.) (Fee, £3.)	Sherardian Professor of Botany, A. G. TANSLEY, M.A.	Th. F. S. 10	Department of Botany (Botanic Garden).	
General Course (Practical Work)	H. BAKER, M.A., and A. R. CLAPHAM, M.A.	Th. F. S. 11–1	,,	
Lectures additional to General Course with Practical Work. (First year Honour Students.)	A. R. CLAPHAM, M.A. ..	T. 9.15–1 ..	,,	
Introductory Course for Forestry Students (Practical Work only) (continued). (Fee £1.)	H. BAKER, M.A.	W. 10–1 (First half of Term) Th. 2–4 (Second half of Term)	,,	W. 18 Jan. Th. 16 Feb.
Elementary Anatomy and Physiology (with Practical Work) (continued). (First year Honour Students.)	The PROFESSOR	W. 9–1	,,	
Distribution of Plants and Vegetation (continued)		W. 12	,,	
Gymnosperms (with Practical Work) (continued)	A. R. CLAPHAM, M.A.	M. 9.15–1	,,	
Protophyta (with Practical Work) ..	,,	W. 9.15–1	,,	
Mycology (with Practical Work) ..	W. H. WILKINS, M.A.	Th. 10 (all day)	,,	
Physiology (with Practical Work) ..	W. O. JAMES, D.Phil.	F. 10 (all day), S. 9–1.	,,	
Genetics (continued) ..	R. SNOW, M.A. ..	M. 12	,,	

§ Names received on Saturday, 14 January, between 9.30 and 1, when the Sherardian Professor will be glad to advise students as to their future work. The Sherardian Professor particularly requests all students entering the Department to see him during these hours, or, if this is impossible, as soon as possible eck of Full Term.

1933 TRINITY TERM

Subject.	Lecturer.	Time.	Place.	Course begins
§BOTANY.				
General Course. (First Year Honour Students and Preliminary Examination in Forestry.) (Fee, £3.)	Sherardian Professor of Botany, A. G. TANSLEY, M.A.	Th. F. S. 10	Botanical Laboratory (Botanic Garden).	
General Course (Practical Work) ..	H. BAKER, M.A., and A. R. CLAPHAM, M.A.	Th. F. S. 11–1	,,	
Field Classes and Laboratory Work. (First Year Forestry Students.)	H. BAKER, M.A.	W. 9 (all day)	,,	
Reproduction of Plants. (Preliminary Examination in Natural Science.) (Fee, £2.)	The PROFESSOR and W. H. WILKINS, M.A.	M. T. W. 10 .. (Second half of Term only)	,,	M. 22 May
Reproduction of Plants (Practical Work)	W. H. WILKINS, M.A. ..	M. T. W. 11–1	,,	
Field Class. (Honour Students.) ..	The PROFESSOR, H. BAKER, M.A., and A. R. CLAPHAM, M.A.	T. 9.15 (all day)	,,	
Practical Work in connexion with Field Class	,,	W. 9.15	,,	
British Flora and Vegetation ..	The PROFESSOR ..	W. 11.45 ..	,,	
Angiosperms (with Practical Work) ..	A. R. CLAPHAM, M.A. ..	M. 9.15 ..	,,	
Mycology (with Practical Work) ..	W. H. WILKINS, M.A. ..	Th. 10 (all day)	,,	
Physiology (with Practical Work) ..	W. O. JAMES, D.Phil. ..	F. 10 (all day), S. 9–1.	,,	
Elementary Course for Students of Agriculture and Geography (with Practical Work) (Fee, £3.)	A. R. CLAPHAM, M.A. ..	Th. F. 10–1 ..	,,	
Soil Science for Students of Botany, with Laboratory and Field Work.	Reader in Soil Science, C. G. T. MORISON, M.A.	W. 9	School of Rural Economy.	

Lecture course and timetable for the academic year 1932/33, showing the range of lectures and practical classes on offer to students in the Department of Botany. Oxford University Herbaria.

but he had forced botany at Oxford on the road to becoming a research-driven discipline by his skills as a teacher and an organizer, particularly through the recognition that research and teaching quality depended as much on the calibre of the staff as on the environment in which they conducted their work.

Post-war botanical teaching

Continual modification of the existing site was an ineffective way of dealing with the inherent problems of how to study and teach botany in the context of the opportunities presented in the early twentieth century.[69] University College, London, and Cambridge were leading the way, showing what could be done as both the range of subjects offered to students and the methods of teaching expanded.[70] Various approaches had been tried at Oxford since the 1830s but had not taken hold, through lack of interest, resources or conviction. A solution had to be found.

The mood of the university and of the incumbent Sherardian professor had changed.[71] The only practical option for the development of botany at Oxford was to vacate the cramped, dysfunctional site at the Botanic Garden. Therefore, if the Garden would not go to the Parks site, the Botany Department would go to the new Science Area alone. Plans were laid and a building designed by the Slade Lecturer in Architecture at the university, Hubert Worthington, in 1939. It was constructed on South Parks Road after the Second World

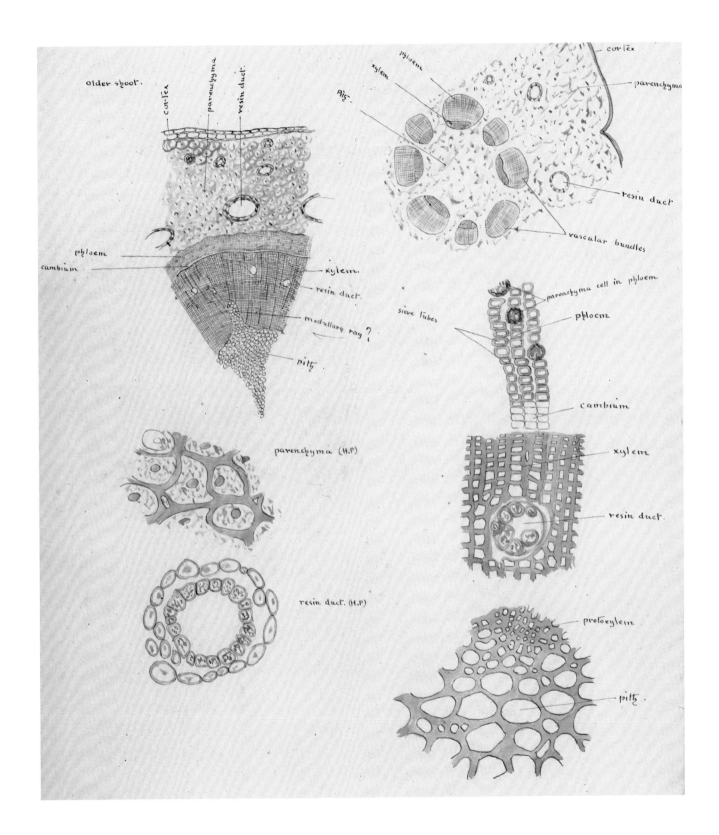

Older shoot.

cortex
parenchyma
resin duct.
Pith
xylem
phloem
cortex
parenchyma
resin duct
vascular bundles

phloem
cambium
xylem.
resin duct.
medullary ray?
pith.

sieve tubes
parenchyma cell in phloem
phloem
cambium
xylem
resin duct.

parenchyma (H.P)

resin duct. (H.P)

protoxylem
pith.

War, and on 8 October 1951 the Botany Department officially moved into its new home, adjacent to the Department of Forestry. Vacant buildings at the Garden were returned to Magdalen College, and the Garden focused its work on horticulture and recreation.

In its new home, the Department of Botany expanded rapidly as new staff were recruited who covered the full range of plant sciences from ecology and systematics, through physiology and biochemistry, to genetics and development. Worthington's design was modified to accommodate the rapidly changing demands of plant sciences in the late twentieth century. Sherardian professors left their marks on the department through the recruiting decisions they made, but by the early 1980s government decisions forced the university to merge the Departments of Botany, Agriculture and Forestry into a single Department of Plant Sciences, where pure and applied sciences were combined. The botany degree, which had been popular among students since the 1950s, was finally combined with zoology in the late 1980s as a biological sciences degree. The Botanic Garden became a regular source of plant material for classes in plant propagation, tropical agriculture and taxonomy classes, while the Garden's collection of hardy spurges (*Euphorbia*) was created specifically for teaching.

In the past 400 years botany at Oxford has perhaps not lived up to expectations for reasons that range from infrastructure, through funding, to the appointments made. However, those who were educated in botany at Oxford, and the botanical research they undertook, have often had great influence through what they did after leaving the university, through their ideas or simply because of the university's institutional prestige.

The next century of plant sciences at Oxford begins with the creation of a new biology department, as the Departments of Plant Sciences and Zoology come together in purpose-built accommodation. Acland's mid-nineteenth-century dream of zoology and plant sciences working together in a shared space is approaching fruition. How plant sciences will fare in a new structure cannot be predicted. Whatever the discipline looks like at Oxford at the end of the twenty-first century, plants will remain central to our lives and the environments in which we live.

Previous page **Early twentieth-century anatomical drawing of sections through the stem of a Scots pine (*Pinus sylvestris*)** by the Oxford entomologist Harry Eltringham. Such observational skills were an essential part of botanical training at the university. Bodleian Library, Sherardian Library of Plant Taxonomy, MS. Sherard 476, f.67r.

Opposite **Twenty-first-century watercolour portrait of thale cress (*Arabidopsis thaliana*)**, a weedy species that has been a model plant since the 1980s, by the botanical artist Barbara McLean. Oxford University Herbaria.

B.M.

Frank White (1927–1994)[72] was born in Sunderland and educated at Cambridge. He was at the time of his death one of the foremost British researchers on the plants and vegetation of Africa. White combined his interests in ecology and systematics in his research on the taxonomy of ebonies and mahoganies, the classification of vegetation and the biology of plant dispersal.

Under his curatorship, the herbaria of the university's then separate Departments of Forestry and Botany were amalgamated, and they became his major research tool. Among many other contributions, White was the author of *The Vegetation of Africa* (1983), which accompanied UNESCO's detailed map of African vegetation categories (devised by him). Data for the map's construction came from White's own extensive African fieldwork, his exploration of the world's herbaria and his collaborations with an extensive network of experts in African plants.

As a believer in the central role of empirical observation in biology, White was a powerful advocate for the fundamental importance of fieldwork. Biological generalizations were valuable only after meticulous observations either in the field or in the herbarium. As a teacher, White inspired generations of undergraduate and postgraduate students with his enthusiasm for rigorous botanical research. Many of his students went on to take up senior taxonomic positions worldwide.

Arthur Harry Church, FRS (1865–1937),[73] the eldest
son of a Plymouth saddler, went to University College Wales,
Aberystwyth, in 1887. Two years later he obtained a B.Sc.
degree in London and won a scholarship to Jesus College,
Oxford, as a mature student in 1891. After graduating with
first-class honours in botany in 1894, he was appointed
demonstrator in the Department of Botany. Church spent
the rest of his career at Oxford, and took early retirement as
a lecturer in 1921. He was particularly affected by the loss of
many friends and students during the First World War.

In *Thalassiophyta* (1919), Church put forward the idea that
land plants arose from marine plants on ocean shores: the
land was invaded by large plants. These ideas were particularly
influential on the work and ideas of the Cambridge botanist
Edred John Henry Corner. Church was an outstanding
botanical illustrator and photographer, although the
publishers of his *Types of Floral Mechanism* (1908) failed to do
his illustrations justice. As a charismatic teacher, Church
learned from his own poor experiences as a student, producing
lectures designed to stimulate the intellect but also to
entertain.

His dry, acidic sense of humour, combined with an
irreverence for university and college politics, led one
Sherardian professor to describe him as a 'genius manqué'.

Arthur George Tansley, FRS (1871–1955),[74] who
was sometimes called 'the Managing Director of British
Ecology', was born in central London to a wealthy family
with strong Christian socialist principles and associations
with the Working Men's College. In 1889 he started to attend
classes at University College, London, where new approaches
to university teaching and research were being devised.[75]

In 1890 he went to Cambridge, taking a first-class degree
in the natural sciences tripos in 1894. During this period,
he maintained an appointment at University College, where
he was working with Francis Wall Oliver, Quain Professor
of Botany. Tansley's launch of the *New Phytologist* in 1902
brought him to botanical prominence. In 1904 he became
the nucleus of a group of botanists who went on to produce
Types of British Vegetation (1911), a systematic description of
British plant communities. The group founded the British
Ecological Society, with Tansley as its first president.

Tansley took up a lectureship at Cambridge in 1907,
which he resigned sixteen years later. He continued to
forge the discipline of ecology, and his *Practical Plant Ecology*
(1923) introduced the subject into schools. Between his
resignation from Cambridge and his appointment as the
tenth Sherardian professor in 1927, he studied and worked
with Sigmund Freud on psychology in Vienna. Following
his retirement from Oxford, Tansley was instrumental in
the formation of the Nature Conservancy in 1949. In his

desire to conserve the British landscape, Tansley took a pragmatic approach, recognizing that the landscapes have been made by people, and that they will continue to change. Consequently, these changes have to be accommodated by conservation practices.

NOTES

PREFACE

1 All conversions are based on the National Archives Currency converter tool (www.nationalarchives.gov.uk/currency) until 2017, with an additional estimation of UK inflation between 2017 and 2020.

CHAPTER 1

1 Dear (2007).
2 Hardy and Totelin (2016).
3 Morton (1981).
4 Collins (2000).
5 Lindberg (2007).
6 Morton (1981).
7 Harris (2015a, pp. 15–19).
8 Gerard (1633, p. 135).
9 Thompson (1934).
10 Gerard (1633, p. 135).
11 Harris (2018).
12 Al-Khalili (2010).
13 Ogilvie (2006).
14 Wilson (2017); Morton (1981).
15 Batey (1986).
16 Ogilvie (2006).
17 Jones (2004); Boulger and McConnell (2004).
18 Arber (1986).
19 Shapin (2018).
20 Bacon (1877, p. 322).
21 Ibid., p. 318.
22 Tinniswood (2019).
23 Harrison (2008).
24 Jardine (2004).
25 Gadd (2014).
26 Syfret (1950).
27 South (1823, p. 374).
28 De Beer (2006, p. 479).
29 Rovelli (2011).
30 Morton (1981).
31 Barash (2018).
32 Livingstone (2013).

33 Brockway (1979).
34 All incumbents of the Sherardian and Sibthorpian chairs have been men. In 2020 the gender imbalance in academic positions in the Department of Plant Sciences remains stark.
35 Brockliss (2016).
36 Jones (1956, p. 273).
37 Chaplin (1920); Allen (1946); Frank (1997).
38 Gutch (1796, p. 335).
39 Worling (2005).
40 Vines and Druce (1914, pp. viii–xiii). For a discussion of previous anniversaries of the Botanic Garden, see Harris (2017a).
41 Daubeny (1853a, p. 13).
42 Gunther (1912, p. 2).
43 Ibid., app. A.
44 'To the Glory of God the best and greatest, to the honour of King Charles, to the use of the university and the State. Henry, Earl Danby 1632'.
45 Gibson (1940, p. 108).
46 Wood (1796, p. 897); Potter (2007, p. 251).
47 Harris (2017a).
48 Sobel (2000).
49 Leith-Ross (1984); Potter (2007).
50 Tradescant (1656).
51 Ibid., p. 139.
52 MacGregor (2001b); Tradescant (1656); MacGregor and Hook (2006).
53 MacGregor (1983).
54 MacGregor (1989).
55 Ibid.
56 Anonymous (1667, p. 321).
57 Grew (1681, p. 183).
58 Lyons (1944, p. 211).
59 MacGregor (1989); Brockliss (2016).
60 Stephens and Browne (1658, 'Preface to the Philobotanick Reader').
61 Vines and Druce (1914, p. xiii).
62 Severn (1839, p. 242).
63 Harris (2018).
64 Hearne (1772, p. 221); Gunther (1912, p. 180).
65 Sorbière (1709, p. 42).
66 Magalotti (1821, p. 262).
67 John Sibthorp's lecture notes, c.1788–94 (Bodleian Libraries, Sherardian Library of Plant Taxonomy, MS. Sherard 219, fol. 19r).

68 Vines (1896).
69 McGurk (2004); MacNamara (1895).
70 Anonymous (1885).
71 Harris (2017a).
72 Allen (2004a).
73 Bobart (1884).
74 Turner (2002); Gunther (1939).
75 Gunther (1925, p. 320); MacGregor (2001a).
76 Letter from Edward Lhwyd to Martin Lister (15 August 1689), transcribed in Gunther (1945, p. 377).

CHAPTER 2

1 When Hetherington et al. (2016) described a fossilized root, *Radix carbonica*, from a 320-million-year-old fossil specimen in Oxford University's botanical collections, Chaffey (2016) quipped that 'Botany at Oxford is not 400 years old!'
2 Harris (2015a, pp. 32–5).
3 Harris (2018).
4 Taiz and Taiz (2017).
5 Clarke and Merlin (2013).
6 Potter (2007).
7 Anonymous (1710).
8 Harris (2017b).
9 Morton (1981).
10 Turner (1586).
11 Morton (1981).
12 Pepys (1854, p. 320).
13 Harris (2018).
14 Evans (1713).
15 Harris (2017a).
16 Lubke and Brink (2004).
17 Harris (2017b).
18 Turner (1835, pp. ix–x).
19 Harris (2017a).
20 Ibid.
21 Harris (2018).
22 Anonymous (1648); Stephens and Browne (1658); 'Catalogus Herbarum ex horto Botanico Oxoniensis anno collectorum 1676', Bobart the younger's autograph manuscript (Bodleian Library, Sherardian Library of Plant Taxonomy, MS. Sherard 32).
23 Harris (2017a); John Sibthorp's lecture notes, c.1788–94 (Bodleian Library, Sherardian Library of Plant Taxonomy, MS. Sherard 219, fols 19–20).

24 Harris (2017a).

25 Heine and Mabberley (1986).

26 Harris (2017a).

27 Daubeny (1853b, p. 15)

28 Clokie (1964); Strugnell (1999).

29 Smith (1816a).

30 John Sibthorp apparently also made an offer for Linnaeus' herbarium (Lack with Mabberley, 1999, pp. 32, 191).

31 Daubeny (1853a).

32 Clokie (1964).

33 Strugnell (1999).

34 Allen (1986).

35 Gunther (1912, p. 149).

36 Henrey (1975).

37 Harris (2007b).

38 Turrill (1938).

39 Burley et al. (2009).

40 Metcalfe (1973).

41 Hillis (1998); Stern (1982); Mills (2004).

42 Burley et al. (2009).

43 Harris (2017a).

44 Dawkins and Field (1978).

45 Savill et al. (2010).

46 Allen (2004d); Clokie (1964).

47 Linnaeus (1737b, dedicatio); Hasselquist (1766, p. 51); Turner (1835, p. xix).

48 Goddard (2004a).

49 Gunther (1904, app. E); Tuckwell (1908); Goddard (2004a).

50 Liebig (1855).

51 Lindley (1836).

52 Jackson and Kell (2004); Clokie (1964).

53 'Memoir of the late Henry Borron Fielding Esqʳᵉ. F.L.S & G.S. of Lancaster', unpublished manuscript (Bodleian Library, Sherardian Library of Plant Taxonomy, MS. Sherard 397, fols 17–18).

54 Ibid., fol. 17.

55 Letter from George Gardner to William Hooker, Kandy, 18 July 1844 (Royal Botanic Gardens, Kew, Directors' Correspondence, 54/167).

56 Harris (2007a).

57 Allen (1986).

CHAPTER 3

1 MacGregor (2018).

2 The observation of the plant by the Bohemian botanist Thaddäus Haenke is regarded as the first time it was seen by a European (Hooker, 1847).

3 Hooker (1847, p. 8).

4 Ibid., p. 10.

5 Ibid., p. 11.

6 *Illustrated London News*, 17 November 1849.

7 Hooker (1847, p. 12).

8 Prance and Arius (1975).

9 Clute (1904, p. 1).

10 Van den Spiegel (1606, p. 79–81).

11 Woodward (1696, p. 12).

12 Graves (1818, pp. 294–5).

13 Hooker (1849, p. 402).

14 Woodward (1696, p. 16).

15 Hooker (1849, pp. 403–4); Baker (1958); Allen (1965); Endersby (2008).

16 Endersby (2008).

17 Donovan (1805, p. 78); see also Brockway (1979).

18 Woodward (1696, p. 16).

19 MacGregor (2018).

20 Raven (1950).

21 Courtney and Davis (2004).

22 Edgington (2016); Turner (1835); Dandy (1958).

23 Edgington (2016); Turner (1835).

24 Letter from Richard Richardson to Johann Dillenius, 25 October 1726, transcribed in Druce and Vines (1907, p. lxxx).

25 Turner (1835, p. 263).

26 Herbarium specimen, Oxford University Herbaria, Mor_II_219_05c.

27 Ibid., Mor_II_244_22b.

28 Turner (1835).

29 Druce and Vines (1907); Turner (1835).

30 Turner (1835, p. 347).

31 Turner (1835).

32 Anonymous (1791).

33 Harris (2019).

34 Shaw (1738). All quotations attributed to Shaw come from this source, unless indicated otherwise.

35 Harris (2019).

36 Anonymous (1791).

37 John Sibthorp's lecture notes, c.1788–94 (Bodleian Library, Sherardian Library of Plant Taxonomy, MS. Sherard 219, fol. 14r);

Harris (2007b).

38 Nicholls (2009).

39 Frick and Stearns (1961, p. 114).

40 Nelson and Elliott (2015).

41 Letter from Mark Catesby, 16 January 1723/24 (Royal Society Archive, Sherard Correspondence, CCLIII 173).

42 Ibid., CCLIII 174.

43 Letter from Mark Catesby to William Sherard, 10 May 1723 (Royal Society Archive, Sherard Correspondence, CCLIII 171).

44 Mark Catesby's specimen and pen-and-ink sketch of *Nelumbo lutea* (Oxford University Herbaria, Sher-1090-10 and Sher-1090-10a).

45 Letter from Mark Catesby to William Sherard, 4 January 1723 (Royal Society Archive, Sherard Letters, CCLIII 168).

46 Letter from Mark Catesby to William Sherard, 10 January 1724/25 (Royal Society Archive, Sherard Letters, CCLIII 184).

47 Harris (2015c); McMillan and Blackwell (2013); McMillan et al. (2013).

48 This description of Sibthorp's journey is précised from Harris (2007b).

49 Lack with Mabberley (1999, p. 32).

50 Ibid., p. 108.

51 Ibid., p. 32.

52 Ibid., p. 42.

53 Ibid., p. 42.

54 Ibid., p. 43.

55 Rix (1975).

56 Lack with Mabberley (1999, p. 73).

57 Ibid., p. 182.

58 Killick et al. (1998).

59 Harris (2007a).

60 Allen (1986, pp. 101–47).

61 Uncatalogued letters from William Bateson to George Druce (Oxford University Herbaria, Druce Archive).

62 Allen (1986, pp. 101–7); Ayres (2012).

63 Druce (1898).

64 Harris (2010).

65 Allen (1986, p. 108).

66 Allen (1986).

67 Harris (2010).

68 Ayres (2012, pp. 83–7).

69 Nelson (2018).

70 Letter from Jacob Bobart the younger to Mary Somerset, Duchess of Badminton, 28 March 1694 (British Library, Sloane

MS. 3343 fol. 37r–v).

71 Ward (1852); Allen (1994).

72 Gunther (1912, p. 131).

73 Strugnell (1999).

74 Bebber et al. (2012); Whitfield (2012).

75 Roberts (2004).

76 Jeffer (1953).

77 Letter from Edward Lhwyd to Martin Lister, 15 August 1689, transcribed in Gunther (1945, pp. 377–8).

78 The genus *Lloydia* was described by the German botanist and ornithologist Heinrich Gottlieb Ludwig Reichenbach, who placed within it Carl Linnaeus' Snowdon lily, *Bulbocodium serotinum* L. The entire genus *Lloydia* is now part of the genus *Gagea*.

79 Nelson and Elliott (2015).

80 Harris (2015c).

81 Frick and Stearns (1961, p. 19).

82 Sterling (2004).

CHAPTER 4

1 Genesis 2: 20.

2 Morton (1981).

3 Vavilov (1992, pp. 337–40).

4 Druce and Vines (1907).

5 Anonymous (1669, p. 935).

6 Morison (1669, pp. 463–99).

7 Genesis 1: 11.

8 Vines (1911).

9 Mandelbrote (2015).

10 Morison (1672, t. 2, t. 11).

11 Morison (1680, preface).

12 Anonymous (1675, p. 327).

13 Hancock (2006); Mandelbrote (2015).

14 Druce and Vines (1907); Vines (1911); Raven (1950); Mandelbrote (2015).

15 Harris (2015b); Turner (1835); Sibbald (1684).

16 Freer (2003, p. 151).

17 Smith (1821, p. 281).

18 Turner (1835, pp. 362–4).

19 Ibid., pp. 326–7.

20 Laird (2015, pp. 143–8); Henrey (1975).

21 Druce and Vines (1907).

22 Hopkins et al. (1998).

23 Mabberley et al. (1995); Wood et al. (2020).

24 Freer (2003, p. 40).

25 Dillenius (1715).

26 Druce and Vines (1907).

27 Downin and Marner (1998).

28 John Sibthorp's lecture notes, c.1788–94 (Bodleian Library, Sherardian Library of Plant Taxonomy, MS. Sherard 219, fol. 579r).

29 Ibid.

30 Smith (1821, p. 245).

31 www.worldfloraonline.org.

32 Harris (2015b).

33 Manuscript of William Sherard's *Pinax* (Bodleian Library, Sherardian Library of Plant Taxonomy, MS. Sherard 44-173).

34 Scotland and Wortley (2003).

35 Mabberley (2000).

36 Anonymous (2015).

37 Harris (2017a).

38 Harris (2007b); Lack with Mabberley (1999).

39 Mulholland et al. (2017).

40 Lack (2015); Mabberley (2017).

41 Henrey (1975); Allen (2010).

42 Ingram (2001, p. 42).

43 Anonymous (1804).

44 Harris (2007b).

45 Allen (2010).

46 Oswald and Preston (2011).

47 Druce (1886, pp. 394–5); Killick et al. (1998, p. 80).

48 Ayres (2012, p. 138).

49 Allen (2004b).

50 Turner (1835, p. 10).

51 Uffenbach (1754, pp. 161–2).

52 Butler (1744, p. 119).

53 Mandelbrote (2015).

54 Vines and Druce (1914, p. lv).

55 Mandelbrote (2004).

56 Boulger and Mabberley (2004); Druce and Vines (1907).

57 Pulteney (1790b, pp. 153–74).

58 Riley (2011).

59 Linnaeus (1737a, p. 80).

CHAPTER 5

1 Morton (1981).

2 Rackham (1945, p. 119).

3 Parkinson (1640, p. 1547).

4 Prest (1981, p. 82); Zirkle (1935, pp. 89–90).

5 Ray (1686, pp. 1–58).

6 Harris (2018).

7 Rea (1665, p. 151).

8 Duthie (1988).

9 Cited in the entry on carnations by Thornton (1807).

10 Ibid.

11 Harris (2018).

12 Shapiro (1969).

13 John Sibthorp's lecture notes, c.1788–94 (Bodleian Library, Sherardian Library of Plant Taxonomy, MS. Sherard 219, fol. 123r).

14 Jardine (2004).

15 Morton (1981).

16 Grew (1682, preface).

17 Darwin (1913); Allan and Schofield (1980).

18 Sharrock (1672, p. 116); Plot (1677, p. 260); Plot and Bobart (1683).

19 Hales (1727).

20 Ray (1686, p. 15); Harris (2017a).

21 Priestley (1772, p. 166).

22 Gest (2000).

23 Harris (2017a).

24 Batey (1986, p. 41).

25 Sharrock (1672, p. 30).

26 Morison (1680, pp. 208–9).

27 Daubeny (1835).

28 Osborn and Mabberley (2014); Juniper et al. (1989).

29 Brockliss (2016, pp. 316–17).

30 Thoday (2007).

31 Liebig (1855, dedication).

32 Gunther (1904).

33 Daubeny (1841).

34 Aulie (1974).

35 Russell (1966).

36 Russell (1942).

37 Gunther (1904, p. 80); Lawes and Gilbert (1895, p. 3).

38 Lawes and Gilbert (1895, p. 3).

39 Vines (1888).

40 Harris (2011b).

41 Morton (1981).

42 Browne (2004).

43 Shull and Stanfield (1939); Zirkle (1951).

44 Clapham (1970).

45 Church (1904); Mabberley (2000).

46 Blackman and Palladino (2004).

47 Harman (2004).

48 Barlow (2018).

49 Dubrovsky and Barlow (2015); Barlow (2015).

50 Harman (2004).

51 Grew (1682, p. 171); Millington was Sedleian Professor of Natural Philosophy, not the Savilian Professor of Astronomy.

52 Morton (1981); Taiz and Taiz (2017).

53 Zirkle (1935); Morton (1981).

54 Bernasconi and Taiz (2002); Blunt (2004).

55 Gärtner (1849), quoted from a translation in Roberts (1929, p. 78).

56 Miller (1768, GEN); Blair (1720, p. 272).

57 Raven (1950, p. 174).

58 Jacob Bobart's autograph list of woody plants (Bodleian Library, Sherardian Library of Plant Taxonomy, MS. Sherard 34, fol. 30v, entry 0476, 'Platanus inter Orientalem et Occidentalem media').

59 Bradley (1718); Walters (1981, pp. 15–29).

60 Leapman (2001).

61 Zirkle (1951).

62 Mabberley (2000).

63 Clarke (2004).

64 Mabberley (2000).

65 Ayres (2012).

66 Robertson and Eardley (1973).

67 Anonymous (1973).

68 Hattersley-Smith (2004).

69 White (1983).

70 Smocovitis (2004).

71 Harman (2004).

72 Osborn and Mabberley (2014); Rendle (1934).

73 Jackson (2015).

74 Howarth (1987).

75 Clapham (1970).

CHAPTER 6

1 Desmond (1998); Brockway (1979); Juma (1989).

2 Brockliss (2016).

3 Harris (2007b).

4 Trott (2009).

5 Harris (2015a, pp. 65–9).

6 Gunther (1912, p. 189).

7 Harris (2017a).

8 Goddard (2004a).

9 Brockliss (2016); Endersby (2008, pp. 271–2); Walters (1981).

10 Brockliss (2016, p. 397).

11 Russell (1966).

12 Clarke and Johnston (2004).

13 Lawes and Gilbert (1895).

14 Goddard (2004b).

15 Osborn and Mabberley (2014).

16 Watson and Osborne (2004).

17 Russell (1966).

18 Waterston and Macmillan Shearer (2006, p. 973).

19 Watson (1939).

20 Douglas (2019).

21 Details in this section draw upon Burley et al. (2009). All quotations are from this source unless indicated otherwise.

22 Smith and Lewis (1991).

23 Allen (1994; 2010).

24 Lightman (2010).

25 Bellamy (1908).

26 Lucas (1979); Hesketh (2009).

27 Letter 2534, Charles Kingsley to C.R. Darwin, 18 November 1859 (Darwin Correspondence project, www.darwinproject.ac.uk).

28 Frank Bellamy's autograph manuscript entitled 'George Claridge Druce: A Memoir of his Botanical Life', c.1935 (Bodleian Library, Sherardian Library of Plant Taxonomy, Druce Archive, uncatalogued); Allen (1986).

29 Harris (2017a).

30 Trewavas and Leaver (2001).

31 Burley et al. (2009) give a detailed account of the involvement of Oxford-based academic foresters with real-world forestry practice.

32 Ramsbottom (1942); Ornduff (1980).

33 Harley and Palladino (2004); Harley (1981).

CHAPTER 7

1 Green (1969); Brockliss (2016).

2 Evans (1713).

3 Harris (2002).

4 Druce (1927, p. 241).

5 Allan and Pannell (2009).

6 Walters (1981).

7 Morton (1981).

8 Harris (2018).

9 Vines and Druce (1914, p. xxiv).

10 Pulteney (1790a, p. 301).

11 Clark (1894, p. 49).

12 Power (1919, p. 119).

13 Ibid., p. 117.

14 Evans (1713).

15 Walters (1981, p. 21).

16 Ibid., pp. 15–29; Egerton (2006); Santer (2009).

17 Physic Garden Committee Memorandum, 7 February 1735 (Bodleian Library, Sherardian Library of Plant Taxonomy, MS. Sherard 1, fol. 5r). From the context of the memorandum, the date is wrong and was corrected later to 7 February 1736.

18 Walters (1981, pp. 30–46); Brockliss (2016, pp. 225–322).

19 Harris (2017a).

20 Brockliss (2016, p. 316).

21 Smith (1816b); Clokie (1964, p. 36).

22 Gascoigne (2004).

23 Glynn (2002).

24 Ibid.

25 Allen (2004c); Walters (1981, pp. 30–5); Allen (1994, p. 10).

26 Walters (1981, p. 36).

27 Boulger and Sherbo (2004); Walters (1981, pp. 36–44).

28 Harris (2007b).

29 583-page manuscript in Sibthorp's hand, extensively revised and annotated by him (Bodleian Library, Sherardian Library of Plant Taxonomy, MS. Sherard 219).

30 John Sibthorp's lecture notes, c.1788–94 (Bodleian Library, Sherardian Library of Plant Taxonomy, MS. Sherard 219, fol. 72).

31 Harris (2011a).

32 John Sibthorp's lecture notes, c.1788–94 (Bodleian Library, Sherardian Library of Plant Taxonomy, MS. Sherard 219, fol. 39).

33 Walters (1981).

34 John Sibthorp's lecture notes, c.1788–94 (Bodleian Library, Sherardian Library of Plant Taxonomy, MS. Sherard 219, fol. 137).

35 Ibid., fol. 143.

36 Ibid., fol. 19.

37 Ibid., fol. 393.

38 Ibid., fol. 19.

39 Ibid., fol. 578.

40 Ibid., fols 570–71.

41 Ibid., fol. 583.

42 Printed fliers advertising lecture courses by Williams, 21 April 1819 and 19 April 1817, uncatalogued (Bodleian Library, Sherardian Library of Plant Taxonomy); Tuckwell (1908, p. 34).

43 Goddard (2004a).

44 Brockliss (2016).

45 Acland (1893).

46 Yanni (2005).

47 Daubeny (1853a, p. 13).

48 Gunther (1912, p. 28).

49 Lawson (1875–76).

50 Gunther (1912, p. 28).

51 Schönland (1888; 1886).

52 Ibid.

53 Lubke and Brink (2004).

54 Gunther (1912, pp. 155–60).

55 Harris (2017a).

56 Walters (1981, pp. 70–72); Bower (1938, p. 52).

57 Bower (1938, p. 30).

58 Ibid., p. 52.

59 Gunther (1967, p. iii); Howarth (1987).

60 Mabberley (2000, p. 26).

61 Mabberley (2000).

62 Vines (1896).

63 Mabberley (2000, p. 68).

64 Harris (2017a).

65 Ayres (2012, p. 126); Walters (1981, pp. 83–94); Boulter (2017).

66 Mabberley (2000).

67 Ayres (2012, pp. 127–32).

68 Allen (1986, pp. 116–18); Ayres (2012, p. 131).

69 Gunther (1916).

70 Walters (1981, pp. 83–94); Boulter (2017).

71 Anonymous (1951).

72 Cronk and Sugden (1994); Angus and Chapman (1996).

73 Mabberley (2004; 2000).

74 Hope-Simpson and Evans (2004); Ayres (2012).

75 Boulter (2017).

REFERENCES

Acland, H.W. (1893), *The Oxford Museum*, George Allen, London.

Al-Khalili, J. (2010), *Pathfinders: The Golden Age of Arabic Science*, Allen Lane, London.

Allan, D.G.C., and R.E. Schofield (1980), *Stephen Hales: Scientist and Philanthropist*, Scolar Press, London.

Allan, E., and J.R. Pannell (2009), 'Rapid Divergence in Physiological and Life-History Traits between Northern and Southern Populations of the British Introduced Neo-Species, *Senecio squalidus*', *Oikos*, vol. 118, pp. 1053–61.

Allen, D.E. (1965), 'Some Further Light on the History of the Vasculum', *Proceedings of the Botanical Society of the British Isles*, vol. 6, pp. 105–9.

Allen, D.E. (1986), *The Botanists: A History of the Botanical Society of the British Isles through a Hundred and Fifty Years*, St Paul's Bibliographies, Winchester.

Allen, D.E. (1994), *The Naturalist in Britain: A Social History*, Princeton University Press, Princeton.

Allen, D.E. (2004a), 'Bobart [Bobert], Jacob, the Elder (*c*.1599–1680)', in *Oxford Dictionary of National Biography*, doi: 10.1093/ref:odnb/2741.

Allen, D.E. (2004b), 'Bobart, Jacob, the Younger (1641–1719)', in *Oxford Dictionary of National Biography*, doi: 10.1093/ref:odnb/2742.

Allen, D.E. (2004c), 'Martyn, John (1699–1768)', in *Oxford Dictionary of National Biography*, doi: 10.1093/ref:odnb/18235.

Allen, D.E. (2004d), 'Sherard, William (1659–1728)', in *Oxford Dictionary of National Biography*, doi: 10.1093/ref:odnb/25355

Allen, D.E. (2010), *Books and Naturalists*, Collins, London.

Allen, P. (1946), 'Medical Education in 17th Century England', *Journal of the History of Medicine and Allied Sciences*, vol. 1, pp. 115–43.

Angus, A., and J.D. Chapman (1996), 'A Tribute to Frank White (5th March 1927 to 12th September 1994)', *Bothalia*, vol. 26, pp. 69–76.

Anonymous (1648), *Catalogus plantarum horti medici Oxoniensis*, Henricus Hall, Oxford.

Anonymous (1667), 'Observables Touching Petrification', *Philosophical Transactions of the Royal Society of London*, vol. 1, pp. 320–21.

Anonymous (1669), 'An Account of Books – I. *Praeludia botanica Roberti Morison Scoti Aberdonensis*. Londini, impensis Jac. Allestry, 1669', *Philosophical Transactions of the Royal Society of London*, vol. 4, pp. 934–5.

Anonymous (1675), 'A proposal to Noblemen, gentlemen and others, who are willing to subscribe towards Dr. Morison's New Universal Herbal, ordering plants according to a new and true method, never published heretofore', *Philosophical Transactions of the Royal Society of London*, vol. 10, pp. 327–8.

Anonymous (1710), 'From my Own Apartment', *The Tatler*, no. 216 (August 25), p. 150.

Anonymous (1791), 'The Life of Thomas Shaw, D.D. Principal of St. Edmund's Hall, Oxford', *European Magazine*, vol. 19, pp. 83–6.

Anonymous (1804), *Monthly Magazine, or, British Register*, vol. 17, p. 346.

Anonymous (1885), 'Jacob Bobart', *Gardeners' Chronicle*, vol. 24, pp. 208–9.

Anonymous (1951), *University of Oxford, Department of Botany: Opening of the New Building by the Lord Rothschild*, University Press, Oxford.

Anonymous (1973), 'Obituary: Professor T.G. Osborn. Botany at Oxford', *The Times* (6 June), p. 20.

Anonymous (2015), 'Sibthorp Medal Awarded', *Oxford Plant Systematics*, no. 21, p. 3.

Arber, A. (1986), *Herbals*, Cambridge University Press, Cambridge.

Aulie, R.P. (1974), 'The Mineral Theory', *Agricultural History*, vol. 48, pp. 369–82.

Ayres, P. (2012), *Shaping Ecology: The Life of Arthur Tansley*, Wiley-Blackwell, Chichester.

Bacon, F. (1877), *The New Atlantis, The Wisdom of the Ancients, The History of King Henry VII and Historical Sketches*, Ward, Lock, and Tyler, London.

Baker, H.G. (1958), 'Origin of the Vasculum', *Proceedings of the Botanical Society of the British Isles*, vol. 3, pp. 41–3.

Barash, D.P. (2018), *Through a Glass Brightly: Using Science to See our Species as We Really Are*, Oxford University Press, Oxford.

Barlow, P.W. (2015), 'The Concept of the Quiescent Centre and How It Found Support from Work with X-Rays. I: Historical Perspectives', *Plant Root*, vol. 9, pp. 43–55.

Barlow, P.W. (2018), 'FAL Clowes, 1921–2016: A Memoir', *Plant Signaling & Behavior*, vol 13, e1274484.

Batey, M. (1986), *Oxford Gardens: The University's Influence on Garden History*, Scolar Press, Aldershot.

Bebber, D.P., M.A. Carine, G. Davidse, D.J. Harris, E.M. Haston, M.G. Penn, S. Cafferty, J.R. Wood and R.W. Scotland (2012), 'Big Hitting Collectors Make Massive and Disproportionate Contribution to the Discovery of Plant Species', *Proceedings of the Royal Society B: Biological Sciences*, vol. 279, pp. 2269–74.

Bellamy, F.A. (1908), *A Historical Account of the Ashmolean Natural History Society of Oxfordshire, 1880–1905*, published by the author, Oxford.

Bernasconi, P., and L. Taiz (2002), 'Sebastian Vaillant's 1717 Lecture on the Structure and Function of Flowers', *Huntia*, vol. 11, pp. 97–128.

Blackman, V.H., and P. Palladino (2004), 'Farmer, Sir John Bretland (1865–1944)', in *Oxford Dictionary of National Biography*, doi: 10.1093/ref:odnb/33082.

Blair, P. (1720), *Botanick Essays*, London, Williams and John Innys.

Blunt, W. (2004), *Linnaeus: The Compleat Naturalist*, Frances Lincoln, London.

Bobart, H.T. (1884), *A Biographical Sketch of Jacob Bobart, of Oxford, Together with an Account of his Two Sons, Jacob and Tilleman*, printed for private circulation, Leicester.

Boulger, G.S., and D.J. Mabberley (2004), 'Dillenius, Johann Jakob (1687–1747)', in *Oxford Dictionary of National Biography*, doi: 10.1093/ref:odnb/7648.

Boulger, G.S., and A. McConnell (2004), 'Lyte, Henry (1529?–1607)', in *Oxford Dictionary of National Biography*, doi: 10.1093/ref:odnb/17301.

Boulger, G.S., and A. Sherbo (2004), 'Martyn, Thomas (1735–1825)', in *Oxford Dictionary of National Biography*, doi: 10.1093/ref:odnb/18239.

Boulter, M. (2017), *Bloomsbury Scientists: Science and Art in the Wake of Darwin*, UCL Press, London.

Bower, F.O. (1938), *Sixty Years of Botany in Britain (1875–1935): Impressions of an Eyewitness*, Macmillan, London.

Bradley, R. (1718), *New Improvements of Planting and Gardening, Both Philosophical and Practical, Explaining the Motion of the Sapp and Generation of Plants*, printed for W. Mears, London.

Brockliss, L.W.B. (2016), *The University of Oxford: A History*, Oxford University Press, Oxford.

Brockway, L.H. (1979), *Science and Colonial Expansion: The Role of the British Royal Botanic Gardens*, Yale University Press, New Haven.

Browne, J. (2004), 'Knight, Thomas Andrew (1759–1838)', in *Oxford Dictionary of National Biography*, doi: 10.1093/ref:odnb/15737.

Burley, J., R.A. Mills, R.A. Plumptre, P.S. Savill, P.J. Wood and H.L. Wright (2009), 'Witness to History: A History of Forestry at Oxford University', *British Scholar*, vol. 1, pp. 236–61.

Butler, S. (1744), *Hudibras, in three parts, written in the time of the late wars: corrected and amended. With large annotations, and a preface, by Zachary Grey*, vol. 1, printed for Robert Owen and William Brien, Dublin.

Chaffey, N. (2016), 'Botany at Oxford University is Not 400 Years Old!', *Botany One* (18 October), www.botany.one/2016/10/botany-oxford-university-not-400-years-old.

Chaplin, A. (1920), 'The History of Medical Education in the Universities of Oxford and Cambridge, 1500–1850', *Proceedings of the Royal Society of Medicine*, vol. 13, pp. 83–107.

Church, A.H. (1904), *On the Relation of Phyllotaxis to Mechanical Laws*, Williams & Norgate, London.

Clapham, A.R. (1970), 'George Robert Sabine Snow, 1897–1969', *Biographical Memoirs of Fellows of the Royal Society*, vol. 16, pp. 498–522.

Clark, A. (1894), *The Life and Times of Anthony Wood, Antiquary, of Oxford, 1632–1695. Vol. III: 1682–1695*, Clarendon Press for the Oxford Historical Society, Oxford.

Clarke, C. (2004), 'Ford, Edmund Brisco (1901–1988)', in *Oxford Dictionary of National Biography*, doi: 10.1093/ref:odnb/40012.

Clarke, E., and A.E. Johnston (2004), 'Gilbert, Sir Joseph Henry (1817–1901)', in *Oxford Dictionary of National Biography*, doi: 10.1093/ref:odnb/33399.

Clarke, R.C., and M.D. Merlin (2013), *Cannabis: Evolution and Ethnobotany*, University of California Press, Berkeley.

Clokie, H.N. (1964), *An Account of the Herbaria of the Department of Botany in the University of Oxford*, Oxford University Press, Oxford.

Clute, W.N. (1904), *The Making of an Herbarium*, Bulletin No. IV, Roger William Park Museum, Providence, Rhode Island.

Collins, M. (2000), *Medieval Herbals: The Illustrative Traditions*, British Library, London, and Toronto University Press, Toronto.

Courtney, W.P., and P. Davis (2004), 'Richardson, Richard (1663–1741)', in *Oxford Dictionary of National Biography*, doi: 10.1093/ref:odnb/23576.

Cronk, Q.C.B., and A.M. Sugden (1994), 'Obituary: Frank White', *The Independent*, 18 October.

Dandy, J.E. (1958), *The Sloane Herbarium: An annotated list of the Horti Sicci composing it with biographical accounts of the principal contributors, based on records compiled by the late James Britten*, Trustees of the British Museum, London.

Darwin, F. (1913), 'Stephen Hales, 1677–1761', in F.W. Oliver (ed.), *Makers of British Botany: A Collection of Biographies by Living Botanists*, Cambridge University Press, Cambridge, pp. 65–83.

Daubeny, C. (1835), 'On the Action of Light upon Plants, and of Plants upon the Atmosphere', *Philosophical Transactions of the Royal Society of London*, vol. 126, pp. 149–75.

Daubeny, C. (1841), *Three Lectures in Agriculture; delivered at Oxford, on July 22nd, and Nov. 25th, 1840, and on Jan. 26th, 1841, in which chemical operation of manures is particularly considered, and the scientific principles explained, upon which their efficacy appears to depend*, John Murray, Oxford.

Daubeny, C. (1853a), *Address to the Members of the University, delivered on May 20, 1853*, Botanic Garden, Oxford.

Daubeny, C. (1853b), *Oxford Botanic Garden, or, A Popular Guide to the Botanic Garden of Oxford*, 2nd edn, Messrs. Parker, Oxford.

Dawkins, H.C., and D.R.B. Field (1978), *A Long-Term Surveillance System for British Woodland Vegetation*, Commonwealth Forestry Institute, Oxford.

Dear, P. (2007), *The Intelligibility of Nature: How Science Makes Sense of the World*, University of Chicago Press, Chicago.

De Beer, E.S. (2006), *The Diary of John Evelyn*, Everyman's Library, London.

Desmond, R. (1998), *Kew: The History of the Royal Botanic Gardens*, Harvill Press, London.

Dillenius, J.J. (1715), 'Dissertatio epistolaris de plantarum propagatione maxime capillarium et muscorum cum iconibus et descriptionibus herbarum aliquota novarum', *Academiae Caesaro-Leopoldinae Naturae Curiosorum Ephemerides, sive, Observationum medico-physicarum*, vols 5–6 (appendix), pp. 45–68.

Donovan, E. (1805), *Instructions for Collecting and Preserving Various Subjects of Natural History; quadrupeds, birds, reptiles, fishes, shells, corals, plants, &c. together with a treatise on the management of insects in their several states; selected from the best authorities*, F.C. and J. Rivington, London.

Douglas, A.E. (2019), 'Sir David Cecil Smith (21 May 1930–29 June 2018)', *Bibliographical Memoirs of Fellows of the Royal Society*, vol. 67, pp. 403–19.

Downin, A., and S. Marner (1998), 'The First Moss to be Collected in Australia? *Leucobryum Candidum* – Collected by William Dampier in 1699', *Journal of Bryology*, vol. 20, pp. 237–40.

Druce, G.C. (1886), *The Flora of Oxfordshire, being a topographical and historical account of the flowering plants and ferns found in the county, with sketches of the progress of Oxfordshire botany during the last three centuries*, Parker & Co., Oxford.

Druce, G.C. (1898), '*Bromus interruptus*', *Journal of Botany*, vol. 34, p. 319.

Druce, G.C. (1927), *The Flora of Oxfordshire: A topographical and historical account of the flowering plants and ferns found in the county; with biographical notices of the botanists who have contributed to Oxfordshire botany during the last four centuries*, Clarendon Press, Oxford.

Druce, G.C., and S.H. Vines (1907), *The Dillenian Herbaria: An Account of the Dillenian Collections in the Herbarium of the University of Oxford*, Clarendon Press, Oxford.

Dubrovsky, J.G., and P.W. Barlow (2015), 'The Origins of the Quiescent Centre Concept', *New Phytologist*, vol. 206, pp. 493–6.

Duthie, R. (1988), *Florists' Flowers and Societies*, Shire Press, Princes Risborough.

Edgington, J. (2016), 'Natural History Books in the Library of Dr Richard Richardson', *Archives of Natural History*, vol. 43, pp. 57–75.

Egerton, F. (2006), 'A History of the Ecological Sciences, Part 20: Richard Bradley, Entrepreneurial Naturalist', *Bulletin of the Ecological Society of America*, vol. 87, pp. 117–27.

Endersby, J. (2008), *Imperial Nature: Joseph Hooker and the Practices of Victorian Science*, University of Chicago Press, Chicago.

Evans, A. (1713), *Vertumnus: An Epistle to Mr. Jacob Bobart, Botany Professor to the University of Oxford, and Keeper of the Physick Garden*, printed by L.L. for Stephen Fletcher Bookseller, Oxford.

Frank, R.G. (1997), 'Medicine', in N. Tyacke, *The History of the University of Oxford*. Vol. IV: *Seventeenth-Century Oxford*, Clarendon Press, Oxford, pp. 505–57.

Freer, S. (2003), *Linnaeus' Philosophia botanica*, Oxford University Press, Oxford.

Frick, G.F., and R.P. Stearns (1961), *Mark Catesby: The Colonial Audubon*, University of Illinois Press, Urbana.

Gadd, I. (2014), *The History of Oxford University Press*. Vol. I: *Beginnings to 1780*, Oxford University Press, Oxford.

Gärtner, C.F. von. (1849), *Versuche und Beobachtungen über die Bastarderzeugung in Pflanzenreich, mit Hinweisung auf die ähnlichen Erscheinungen im Thierreiche*, K.F. Hering, Stuttgart.

Gascoigne, J. (2004), 'Banks, Sir Joseph (1743–1820)', in *Oxford Dictionary of National Biography*, doi: 10.1093/ref:odnb/1300.

Gerard, J. (1633), *The Herball, or, General Historie of Plantes: gathered by John Gerarde of London Master in Chirurgerie, very much enlarged and amended by Thomas Johnson citizen and apothecarye*, Adam Islip, Joice Newton & Richard Whitakers, London.

Gest, H. (2000), 'Bicentenary homage to Dr Jan Ingen-Housz, MD (1730–1799), pioneer of photosynthesis research', *Photosynthesis Research*, vol. 63, pp. 183–90.

Gibson, S. (1940), 'Brian Twyne', *Oxoniensia*, vol. 5, pp. 94–114.

Glynn, L.B. (2002), 'Israel Lyons: a short but starry career. The life of an eighteenth-century Jewish botanist and astronomer', *Notes and Records of the Royal Society of London*, vol. 56, pp. 275–305.

Goddard, N. (2004a), 'Daubeny, Charles Giles Bridle (1795–1867)', in *Oxford Dictionary of National Biography*, doi: 10.1093/ref:odnb/7187.

Goddard, N. (2004b), 'Warington, Robert (1838–1907)', in *Oxford Dictionary of National Biography*, doi: 10.1093/ref:odnb/36744.

Graves, G. (1818), *The Naturalist's Pocket-Book, or Tourist's Companion: Being a brief introduction to the different branches of natural history, with approved methods for collecting and preserving the various productions of nature*, Longman, Hurst, Rees, Orme & Brown, London.

Green, V.H.H. (1969), *The Universities*, Penguin, Harmondsworth.

Grew, N. (1681), *Musaeum regalis societatis, or, A catalogue & description of the natural and artificial rarities belonging to the Royal Society and preserved at Gresham Colledge*, printed by W. Rawlins for the author, London.

Grew, N. (1682), *The Anatomy of Plants, with an idea of a philosophical history of plants, and several other lectures, read before the Royal Society*, printed by W. Rawlins for the author, London.

Gunther, A.E. (1967), *Robert T. Gunther: A Pioneer in the History of Science, 1869–1940*, printed for subscribers, Oxford.

Gunther, R.T. (1904), *A History of the Daubeny Laboratory, Magdalen College, Oxford: to which is appended a list of the writings of Dr. Daubeny, and a register of the names of persons who have attended the chemical lectures of Dr. Daubeny from 1822 to 1867*, Henry Frowde, London.

Gunther, R.T. (1912), *Oxford Gardens Based upon Daubeny's Popular Guide of the Physick Garden of Oxford: with notes on the gardens of the colleges and on the University Park*, Parker & Son, Oxford.

Gunther, R.T. (1916), *The Daubeny Laboratory Register 1904–1915: with notes on the teaching of natural philosophy and with lists of scientific researches carried out by members of Magdalen College, Oxford*, printed for subscribers at the University Press, Oxford.

Gunther, R.T. (1925), *Early Science in Oxford*. Vol. III, part I: *The Biological Sciences*; part II: *The Biological Collections*, printed for subscribers, Oxford.

Gunther, R.T. (1939), *Early Science in Oxford*. Vol. XII: *Dr Plot and the Correspondence of the Philosophical Society of Oxford*, printed for subscribers, Oxford.

Gunther, R.T. (1945), *Early Science in Oxford*. Vol. XIV: *Life and Letters of Edward Lhwyd*, printed for subscribers, Oxford.

Gutch, J. (ed.) (1796), *The History and Antiquities of the University of Oxford in Two Books: by Anthony à Wood, M.A. of Merton College*, vol. 2, printed for the editor, Oxford.

Hales, S. (1727), *Vegetable Staticks, or, an Account of Some Statical Experiments on the Sap in Vegetables; also, a specimen of an attempt to analyse the air*, printed for W. and J. Innys, and T. Woodward, London.

Hancock, A. (2006), 'Robert Morison, the First Professor of Botany at Oxford', *Oxford Plant Systematics*, vol. 13, pp. 14–15.

Hardy, G., and L. Totelin (2016), *Ancient Botany*, Routledge, London.

Harley, J.L. (1981), 'Geoffrey Emett Blackman, 17 April 1903–8 February 1980', *Biographical Memoirs of Fellows of the Royal Society*, vol. 27, pp. 45–82.

Harley, J.L., and P. Palladino (2004), 'Blackman, Geoffrey Emett (1903–1980)', in *Oxford Dictionary of National Biography*, doi: 10.1093/ref:odnb/30823.

Harman, O.S. (2004), *The Man who Invented the Chromosome: A Life of Cyril Darlington*, Harvard University Press, Cambridge, MA.

Harris, S.A. (2002), 'Introduction of Oxford Ragwort, *Senecio squalidus* L. (Asteraceae), to the United Kingdom', *Watsonia*, vol. 24, pp. 31–43.

Harris, S.A. (2007a), 'Druce and Oxford University Herbaria', *Oxford Plant Systematics*, vol. 14, pp. 12–13.

Harris, S.A. (2007b), *The Magnificent* Flora Graeca: *How the Mediterranean Came to the English Garden*, Bodleian Library, Oxford.

Harris, S.A. (2010), 'The Trower Collection: Botanical Watercolours of an Edwardian Lady', *Journal of the History of Collections*, vol. 22, pp. 115–28.

Harris, S.A. (2011a), 'John Sibthorp: Teacher of Botany', *Oxford Plant Systematics*, vol. 17, pp. 16–17.

Harris, S.A. (2011b), *Planting Paradise:*

Cultivating the Garden 1501–1900, Bodleian Library, Oxford.

Harris, S.A. (2015a), *What Have Plants Ever Done for Us? Western Civilization in Fifty Plants*, Bodleian Library, Oxford.

Harris, S.A. (2015b), 'William Sherard: His Herbarium and his *Pinax*', *Oxford Plant Systematics*, vol. 21, pp. 13–15.

Harris, S.A. (2015c), 'The Plant Collections of Mark Catesby in Oxford', in E.C. Nelson and D.J. Elliott (eds), *The Curious Mister Catesby: A 'Truly Ingenious' Naturalist Explores New Worlds*, University of Georgia Press, Athens, pp. 173–88.

Harris, S.A. (2017a), *Oxford Botanic Garden and Arboretum: A Brief History*, Bodleian Library, Oxford.

Harris, S.A. (2017b), 'Herbaria in the Botanic Garden', *Oxford Plant Systematics*, vol. 23, pp. 8–9.

Harris, S.A. (2018), 'Seventeenth-Century Plant Lists and Herbarium Collections: A Case Study from the Oxford Physic Garden', *Journal of the History of Collections*, vol. 30, pp. 1–14.

Harris, S.A. (2019), 'Thomas Shaw's Eighteenth-Century Levantine and Barbary Plants', *Oxford Plant Systematics*, vol. 25, pp. 10–11.

Harrison, P. (2008), 'Religion, the Royal Society, and the Rise of Science', *Theology and Science*, vol. 6, pp. 255–71.

Hasselquist, F. (1766), *Voyages and Travels in the Levant in the Years 1749, 50, 51, 52*, L. Davis and C. Reymers, London.

Hattersley-Smith, G. (2004), 'Polunin, Nicholas Vladimir (1909–1997)', in *Oxford Dictionary of National Biography*, doi: 10.1093/ref:odnb/68834.

Hearne, T. (1772), *The Life of Anthony à Wood from the Year 1632 to 1672*, printed for J. and J. Fletcher in the Turl, and J. Pote, Eton.

Heine, H., and D.J. Mabberley (1986), 'An Oxford Waterlily', *Kew Magazine*, vol. 3, pp. 167–75.

Henrey, B. (1975), *British Botanical and Horticultural Literature before 1800: Comprising a History and Bibliography of Botanical and Horticultural Books Printed in England, Scotland, and Ireland from the Earliest Times until 1800*, Oxford University Press, Oxford.

Hesketh, I. (2009), *Of Apes and Ancestors: Evolution, Christianity, and the Oxford Debate*, University of Toronto Press, Toronto.

Hetherington, A.J., J.G. Dubrovsky and J. Dolan (2016), 'Unique Cellular Organization in the Oldest Root Meristem', *Current Biology*, vol. 26, pp. 1629–33.

Hillis, T. (1998), 'Mary Margaret Chattaway (1899–1997)', *IAWA Journal*, vol. 19, pp. 239–40.

Hooker, W.J. (1847), 'Victoria Regia. Victoria Water-Lily', *Curtis's Botanical Magazine*, vol. 73, tab. 4275–8.

Hooker, W.J. (1849), 'Botany', in J.F.W. Herschel, *A Manual of Scientific Enquiry: Prepared for the Use of Officers in Her Majesty's Navy and Travellers in General*, John Murray, London.

Hope-Simpson, J.F., and D.E. Evans (2004), 'Tansley, Sir Arthur George (1871–1955)', in *Oxford Dictionary of National Biography*, doi: 10.1093/ref:odnb/36415.

Hopkins, H.C.F., C.R. Huxley, C.M. Pannell, G.T. Prance and F. White (1998), *The Biological Monograph: The Importance of Field Studies and Functional Syndromes for Taxonomy and Evolution of Tropical Plants*, Royal Botanic Gardens Kew, London.

Howarth, J. (1987), 'Science Education in Late-Victorian Oxford: A Curious Case of Failure?' *English Historical Review*, vol. 102, pp. 334–71.

Ingram, J.A. (2001), *Elysium Britannicum, or, The Royal Gardens*, University of Pennsylvania Press, Philadelphia.

Jackson, B.D., and P.E. Kell (2004), 'Fielding, Henry Borron (1805–1851)', in *Oxford Dictionary of National Biography*, doi: 10.1093/ref:odnb/9401.

Jackson, M.B. (2015), 'One Hundred and Twenty-Five Years of the Annals of Botany. Part 1: The First 50 Years (1887–1936)', *Annals of Botany*, vol. 115, pp. 1–18.

Jardine, L. (2004), *The Curious Life of Robert Hooke: The Man Who Measured London*, Harper Perennial, London.

Jeffer, R.H. (1953), 'Edward Morgan and the Westminster Physic Garden', *Proceedings of the Linnean Society London*, vol. 164, pp. 102–33.

Jones, W.H.S. (trans.) (1956), *Pliny: Natural History*, books XXIV–XXVII, Harvard University Press, Cambridge, MA.

Jones, W.R.D. (2004), 'Turner, William (1509/10–1568)', in *Oxford Dictionary of National Biography*, doi: 10.1093/ref:odnb/27874.

Juma, C. (1989), *The Gene Hunters: Biotechnology and the Scramble for Seeds*, Princeton University Press, Princeton.

Juniper, B.E., D.M. Joel and R.J. Robins (1989), *The Carnivorous Plants*, Academic Press, London.

Killick, J., R. Perry and S. Woodell (1998), *The Flora of Oxfordshire*, Pisces Publications, Newbury.

Lack, H.W. (2015), *The Bauers, Joseph, Franz and Ferdinand: Masters of Botanical Illustration. An Illustrated Biography*, Prestel, Munich, London and New York.

Lack, H.W., with D.J. Mabberley (1999), *The Flora Graeca Story: Sibthorp, Bauer, and Hawkins in the Levant*, Oxford University Press, Oxford.

Laird, M. (2015), *A Natural History of English Gardening 1650–1800*, Yale University Press, New Haven.

Lawes, J.B., and J.H. Gilbert (1895), *The Rothamsted Experiments: being an account of some of the results of the agricultural investigations, conducted at Rothamsted in the field, the feeding shed, and the laboratory over a period of fifty years*, William Blackwood and Sons, Edinburgh and London.

Lawson, M. (1875–76), Internal papers for the University Council by Marmaduke Lawson, 20 November 1875 and 14 February 1876, Sherardian Library of Plant Taxonomy, Bodleian Library.

Leapman, M. (2001), *The Ingenious Mr Fairchild: The Forgotten Father of the Flower Garden*, St Martin's Press, New York.

Leith-Ross, P. (1984), *The John Tradescants: Gardeners to the Rose and Lily Queen*, Peter Owen, London.

Liebig, J. von. (1855), *Principles of Agricultural Chemistry: With special reference to the late researches made in England*, John Wiley, New York.

Lightman, B. (2010), *Victorian Popularizers of Science: Designing Nature for New Audiences*, University of Chicago Press, Chicago and London.

Lindberg, D.C. (2007), *The Beginnings of Western Science: The European Scientific Tradition in Philosophical, Religious, and Institutional Context, Prehistory to A.D. 1450*. University of Chicago Press, Chicago and London.

Lindley, J. (1836), '*Daubenya aurea*: Gold Daubenya', *Edwards' Botanical Register*, vol. 21, tab. 1813.

Linnaeus, C. (1737a), *Critica botanica in qua nomina plantarum generica, specifica, & variantia examini subjiciuntur, selectiora confirmantur, indigna rejiciuntur; simulque doctrina circa denominationem plantarum traditur. Seu Fundamentorum botanicorum pars IV. Accedit Johannis Browallii De necessitate historiae naturalis discursus*, Apud Conradum Wishoff, Lugduni Batavorum.

Linnaeus, C. (1737b), *Hortus Cliffortianus*, Amstelaedami.

Livingstone, D.N. (2013), *Putting Science in Its Place: Geographies of Scientific Knowledge*. University of Chicago Press, Chicago.

Lubke, R., and E. Brink (2004), 'One Hundred Years of Botany at Rhodes University', *South African Journal of Science*, vol. 100, pp. 609–14.

Lucas, J.R. (1979), 'Wilberforce and Huxley: A Legendary Encounter', *Historical Journal*, vol. 22, pp. 313–30.

Lyons, H. (1944), *The Royal Society 1660–1940: A History of Its Administration Under Its Charters*, Cambridge University Press, Cambridge.

Mabberley, D.J. (2000), *Arthur Harry Church: The Anatomy of Flowers*, Merrell and the Natural History Museum, London.

Mabberley, D.J. (2004), 'Church, Arthur Harry (1865–1937)', in *Oxford Dictionary of National Biography*, doi: 10.1093/ref:odnb/38462.

Mabberley, D.J. (2017), *Painting by Numbers: The Life and Art of Ferdinand Bauer*, NewSouth, Sydney.

Mabberley, D.J., C.M. Pannell and A.M. Sing (1995), *Meliaceae*, published for Foundation Flora Malesiana by Rijksherbarium/Hortus Botanicus, Leiden.

MacGregor, A. (1983), *Tradescant's Rarities: Essays on the Foundation of the Ashmolean Museum, 1683, with a Catalogue of the Surviving Early Collections*, Clarendon Press, Oxford.

MacGregor, A. (1989), '"A Magazin of All Manner of Inventions": Museums in the Quest for "Salomon's House" in Seventeenth-Century England', *Journal of the History of Collections*, vol. 1, pp. 207–12.

MacGregor, A. (2001a), 'The Ashmolean as a Museum of Natural History, 1683–1860', *Journal of the History of Collections*, vol. 13, pp. 125–44.

MacGregor, A. (2001b), *The Ashmolean Museum: A Brief History of the Institution and its Collections*, Ashmolean Museum, Oxford.

MacGregor, A. (2018), *Naturalists in the Field: Collecting, Recording and Preserving the Natural World from the Fifteenth to the Twenty-First Century*, Brill, Leiden.

MacGregor, A., and M. Hook (2006), *Manuscript Catalogues of the Early Museum Collections. Part II: The Vice-Chancellor's Consolidated Catalogue 1695*, Archaeopress, Oxford.

McGurk, J.J.N. (2004), 'Danvers, Henry, Earl of Danby (1573–1644)', in *Oxford Dictionary of National Biography*, doi: 10.1093/ref:odnb/7133.

McMillan, P.D., and A.H. Blackwell (2013), 'The Vascular Plants Collected by Mark Catesby in South Carolina: Combining the Sloane and Oxford Herbaria', *Phytoneuron*, vol. 2013-73, pp. 1–32.

McMillan, P.D., A.H. Blackwell, C. Blackwell and M.A. Spencer (2013), 'The Vascular Plants in the Mark Catesby Collection at the Sloane Herbarium, with Notes on their Taxonomic and Ecological Significance', *Phytoneuron*, vol. 2013-7, pp. 1–37.

MacNamara, F.N. (1895), *Memorials of the Danvers Family (of Dauntsey and Culworth)*, Hardy & Page, London.

Magalotti, L. (1821), *Travels of Cosmo the Third, Grand Duke of Tuscany, through England, during the Reign of King Charles the Second (1669)*, J. Mawman, London.

Mandelbrote, S. (2004), 'Morison, Robert (1620–1683)', in *Oxford Dictionary of National Biography*, doi: 10.1093/ref:odnb/19275.

Mandelbrote, S. (2015), 'The Publication and Illustration of Robert Morison's *Plantarum historiae universalis Oxoniensis*', *Huntington Library Quarterly*, vol. 78, pp. 349–79.

Metcalfe, C.R. (1973), 'Metcalfe and Chalk's Anatomy of the Dicotyledons and its Revision', *Taxon*, vol. 22, pp. 659–68.

Miller, P. (1768), *The Gardeners Dictionary: containing the best and newest methods pf cultivating and improving the kitchen, fruit, flower garden and nursery*, 8th edn, printed for the author, London.

Mills, R. (2004), '100 Years of Forestry Information from Oxford', *SCONUL Focus*, vol. 32, pp. 34–9.

Morison, R. (1669), *Hortus regius Blesensis auctus: cum notulis durationis & charactismis plantarum tam additarum, quam non scriptarum; item plantarum in eodem horto regio Blesensi aucto contentarum, nemini hucusque scriptarum, brevis & succincta delineatio. Quibus accessere observationes generaliores (plantarum in eodem horto regio Blesensi aucto contentarum) rei herbariae studiosis valde necessariae, & cognitu perutiles. Praeludiorum botanicorum pars prior*, Typis Tho. Roycroft, impensis Jacobi Allestry, Londini.

Morison, R. (1672), *Plantarum umbelliferarum distributio nova, per tabulas cognationis et affinitatis ex Libro Naturae observata & detecta*, Theatro Sheldoniano, Oxonii.

Morison, R. (1680), *Plantarum historiae universalis Oxoniensis pars secunda seu Herbarum distributio nova, per tabulas cognationis & affinitatis ex Libro Naturae observata & detecta*, e Theatro Sheldoniano, Oxonii.

Morton, A.G. (1981), *History of Botanical Science: An Account of the Development of Botany from Ancient Times to the Present Day*, Academic Press, London.

Mulholland, R., D. Howell, A. Beeby, C.E. Nicholson and K. Domoney (2017), 'Identifying Eighteenth Century Pigments at the Bodleian Library Using *in situ* Raman Spectroscopy, XRF and Hyperspectral Imaging', *Heritage Science*, vol. 5, art. 43.

Nelson, E.C. (2018), 'From Tubs to Flying Boats: Episodes in Transporting Living Plants', in A. Macgregor (ed.), *Naturalists in the Field: Collecting, Recording and Preserving the Natural World from the Fifteenth to the Twenty-First Century*, Brill, Leiden, pp. 578–606.

Nelson, E.C., and D.J. Elliott (2015), *The Curious Mister Catesby: A 'Truly Ingenious' Naturalist Explores New Worlds*, University of Georgia Press, Athens, and London.

Nicholls, S. (2009), *Paradise Found: Nature in America at the Time of Discovery*, University of Chicago Press, Chicago and London.

Ogilvie, B.W. (2006), *The Science of Describing: Natural History in Renaissance Europe*, University of Chicago Press, Chicago and London.

Ornduff, R. (1980), 'Joseph Burtt Davy: A Decade in California', *Madroño*, vol. 27, pp. 171–6.

Osborn, T.G.B., and D.J. Mabberley (2014), 'Vines, Sydney Howard (1849–1934)', in *Oxford Dictionary of National Biography*, doi: 10.1093/ref:odnb/36663.

Oswald, P.H., and C.D. Preston (2011), *John Ray's Cambridge Catalogue (1660)*, Ray Society, London.

Parkinson, J. (1640), *Theatrum botanicum: The Theatre of Plants*, printed by Tho. Cotes, London.

Pepys, S. (1854), *Diary and Correspondence of Samuel Pepys, F.R.S., the Diary Deciphered by J. Smith, with a Life and Notes by Richard Lord Braybrooke*, vol. 2, Henry Colburn, London.

Plot, R. (1677), *The Natural History of Oxford-shire, being an Essay toward the Natural History of England*, printed at the Theater, Oxford.

Plot, R., and J. Bobart (1683), 'A discourse concerning the effects of the great frost, on trees and other plants Anno 1683. drawn from the answers to some Queries sent into divers Countries by Dr. Rob Plot S.R.S., and from several observations made at Oxford, by the skilful botanist Mr. Jacob Bobart', *Philosophical Transactions of the Royal Society of London*, vol. 14, pp. 766–79.

Potter, J. (2007), *Strange Blooms: The Curious Lives and Adventures of the John Tradescants*, Atlantic Books, London.

Power, D. (1919), 'The Oxford Physic Garden', *Annals of Medical History*, vol. 2, pp. 109–25.

Prance, G.T., and A.R. Arius (1975), 'A Study of the Floral Biology of *Victoria amazonica* (Poepp.) Sowerby (Nymphaeaceae)', *Acta Amazonica*, vol. 5, pp. 109–39.

Prest, J. (1981), *The Garden of Eden: The Botanic Garden and the Re-creation of Paradise*, Yale University Press, New Haven and London.

Priestley, J. (1772), 'Observations on Different Kinds of Air', *Philosophical Transactions of the Royal Society of London*, vol. 62, pp. 147–224.

Pulteney, R. (1790a), *Historical and Biographical Sketches of the Progress of Botany in England, from its origin to the introduction of the Linnaean system*, vol. 1, printed for T. Cadell, in the Strand, London.

Pulteney, R. (1790b), *Historical and Biographical Sketches of the Progress of Botany in England, from its origin to the introduction of the Linnaean system*, vol. 2, printed for T. Cadell, in the Strand, London.

Rackham, H. (trans.) (1945), *Pliny: Natural History*, books XII–XVI, Harvard University Press, Cambridge, MA.

Ramsbottom, J. (1942), 'Obituary. Dr. Joseph Burtt Davy (1870–1940)', *Proceedings of the Linnean Society of London*, vol. 153, pp. 291–3.

Raven, C.E. (1950), *John Ray: Naturalist*, Cambridge University Press, Cambridge.

Ray, J. (1686), *Historia plantarum*, Typis Mariae Clark: prostant apud Henricum Faithorne & Joannem Kersey, Londini.

Rea, J. (1665), *Flora, seu, De florum cultura, or, A complete florilege, furnished with all requisites belonging to a florist*, Richard Marriott, London.

Rendle, A.B. (1934), 'Sydney Howard Vines. 1849–1934', *Obituary Notices of Fellows of the Royal Society*, vol. 1, pp. 185–8.

Riley, M. (2011), 'Procurers of Plants and Encouragers of Gardening: William and James Sherard and Charles du Bois, case studies in late seventeenth- and early eighteenth-century botanical and horticultural patronage', Ph.D. thesis, University of Buckingham.

Rix, E.M. (1975), 'Notes on *Fritillaria* (Liliaceae) in the Eastern Mediterranean Region, III', *Kew Bulletin*, no. 30, pp. 153–62.

Roberts, B.F. (2004), 'Lhuyd [Lhwyd; *formerly* Lloyd], Edward (1659/60?–1709)', in *Oxford Dictionary of National Biography*, doi: 10.1093/ref:odnb/16633.

Roberts, H.F. (1929), *Plant Hybridization before Mendel*, Princeton University Press, Princeton.

Robertson, R.N., and C.M. Eardley (1973), 'Theodore George Bentley Osborn, D.Sc., M.A., F.L.S. 2.x.1887–3.vi.1973', *Transactions of the Royal Society of South Australia*, vol. 97, pp. 317–20.

Rovelli, C. (2011), *Anaximander*, Westholme, Yardley, PA.

Russell, E.J. (1942), 'Rothamsted and Its Experimental Station', *Agricultural History*, vol. 16, pp. 161–83.

Russell, E.J. (1966), *A History of Agricultural Science in Great Britain, 1620–1954*, George Allen and Unwin, London.

Santer, M. (2009), 'Richard Bradley: A Unified, Living Agent Theory of the Cause of Infectious Diseases of Plants, Animals, and Humans in the First Decades of the 18th Century', *Perspectives in Biology and Medicine*, vol. 52, pp. 566–78.

Savill, P.S., C.M. Perrins, K.J. Kirby and N. Fisher (2010), *Wytham Woods: Oxford's Ecological Laboratory*, Oxford University Press, Oxford.

Schönland, S. (1886), 'Der botanische Garten, das botanische Institut, das botanische Museum, die Herbarien und die botanische Bibliothek der Universität Oxford', *Botanisches Centralbatt*, vol. 25, pp. 187–93.

Schönland, S. (1888), 'The Botanical Laboratory at Oxford', *Botanical Gazette*, vol. 13, pp. 221–4.

Scotland, R.W., and A.H. Wortley (2003), 'How Many Species of Seed Plant Are There?' *Taxon*, vol. 52, pp. 101–4.

Severn, C. (1839), *Diary of the Rev. John Ward, A.M., Vicar of Stratford-upon-Avon, extending from 1648 to 1679*, Henry Colburn, London.

Shapin, S. (2018), *The Scientific Revolution*, University of Chicago Press, Chicago and London.

Shapiro, B.J. (1969), *John Wilkins, 1614–1672: An Intellectual Biography*, University of California Press, Berkeley.

Sharrock, R. (1672), *The History of the Propagation & Improvement of Vegetables by the Concurrence of Art and Nature*, printed by W. Hall, for Ric. Davis, Oxford.

Shaw, T. (1738), 'Specimen Phytographiae Africanae &c. or a Catalogue of some of the rarer plants of Barbary, Egypt and Arabia', in *Travels, or Observations Relating to Several Parts of Barbary and the Levant*, printed at the Theatre, Oxford.

Shull, C.A., and J.F. Stanfield (1939), 'Thomas Andrew Knight in Memoriam', *Plant Physiology*, vol. 14, pp. 1–8.

Sibbald, R. (1684), *Scotland Illustrated, or, An Essay of Natural History*, printed by J.K., J.S., and J.C., Edinburgh.

Smith, D., and D. Lewis (1991), 'Professor J.L. Harley', *New Phytologist*, vol. 119, pp. 5–7.

Smith, J.E. (1816a), 'Sherard, William', in A. Rees (ed.), *The New Cyclopaedia*, vol. 32, part 2, Longman, Hurst, Rees, Orme, and Brown, London.

Smith, J.E. (1816b), 'Sibthorpia', in A. Rees (ed.), *The New Cyclopaedia*, vol. 32, part 2, Longman, Hurst, Rees, Orme, and Brown, London.

Smith, J.E. (1821), *A Selection of the Correspondence of Linnaeus and Other Naturalists, from the Original Manuscripts*, vol. 2, Longman, Hurst, Rees, Orme, and Brown, London.

Smocovitis, V.B. (2004), 'Darlington, Cyril Dean (1903–1981)', in *Oxford Dictionary of National Biography*, doi: 10.1093/ref:odnb/31000.

Sobel, D. (2000), *Galileo's Daughter*, Fourth Estate, London.

Sorbière, S. (1709), *A Voyage to England: containing many things relating to the state of learning, religion, and other curiosities of that Kingdom. As also, observations on the same voyage, by Dr. Thomas Sprat, Lord Bishop of Rochester. With a letter of Monsieur Sorbière's, concerning the war between England and Holland in 1652: to all which is prefix'd his life writ by M. Graverol*, J. Woodward, London.

South, R. (1823), *Sermons Preached upon Several Occasions*, vol. 1, Clarendon Press, Oxford.

Stephens, P. and W. Browne (1658), *Catalogus horti botanici Oxoniensis*, Typis Gulielmi Hall, Oxonii.

Sterling, K.B. (2004), 'Sibthorp, John (1758–1796)', in *Oxford Dictionary of National Biography*, doi: 10.1093/ref:odnb/25509.

Stern, W.L. (1982), 'Highlights in the Early History of the International Association of Wood Anatomists', in P. Baas (ed.), *New Perspectives in Wood Anatomy*, M. Nijhoff, The Hague, and W. Junk, London, pp. 1–21.

Strugnell, A. (1999), 'The History of the Daubeny Herbarium (FHO): 75th Anniversary', *Oxford Plant Systematics*, vol. 7, pp. 14–16.

Syfret, R.H. (1950), 'Some Early Reactions to the Royal Society', *Notes and Records: The Royal Society Journal of the History of Science*, vol. 7, pp. 207–58.

Taiz, L., and L. Taiz (2017), *The Discovery and Denial of Sex in Plants*, Oxford University Press, Oxford.

Thoday, P. (2007), *Two Blades of Grass: The Story of Cultivation*, Thoday Associates, Corsham.

Thompson, C.J.S. (1934), *The Mystic Mandrake*, Rider & Co., London.

Thornton, R.J. (1807), *New Illustration of the Sexual System of Carolus von Linnaeus: and the temple of Flora, or garden of nature*, published for author, London.

Tinniswood, A. (2019), *The Royal Society and the Invention of Modern Science*, Head of Zeus, London.

Tradescant, J. (1656), *Musaeum Tradescantianum, or, A Collection of Rarities Preserved at South-Lambeth neer London*, printed by John Grismond, London.

Trewavas, A., and C.J. Leaver (2001), 'Is Opposition to GM Crops Science or Politics? An Investigation into the Arguments that GM Crops Pose a Particular Threat to the Environment', *EMBO Reports*, vol. 2, pp. 455–9.

Trott, M. (2009), 'The Sibthorps of Canwick: The Rise and Fall of a Dynasty', in S. Brook, A. Walker and R. Wheeler (eds), *Lincoln Connections: Aspects of City and County since 1700*, Society of Lincolnshire Heritage and Archaeology, Lincoln, pp. 43–58.

Tuckwell, W. (1908), *Reminiscences of Oxford*, E.P. Dutton, New York.

Turner, A.J. (2002), 'Plot, Robert (bap. 1640, d. 1696)', in *Oxford Dictionary of National Biography*, doi: 10.1093/ref:odnb/22385.

Turner, D. (1835), *Extracts from the Literary and Scientific Correspondence of Richard Richardson, M.D., F.R.S., of Bierley, Yorkshire*, printed by Charles Sloman, Yarmouth.

Turner, W. (1586), *The seconde part of William Turners Herball wherein are conteyned the names of herbes in Greke, Latine, Duche, Frenche and in the Apothecaries Latin and somtyne in Italiane, with the vertues of the same herbes with diverse confutationes of no smalle errours that men of no small learning have committed in the intreating of herbes of late yeares*, Arnold Birchaman, London.

Turrill, W.B. (1938), 'A Contribution to the Botany of Athos Peninsula', *Bulletin of Miscellaneous Information*, vol. 1937, pp. 197–273.

Uffenbach, Z.C. von. (1754), *Merkwürdige Reisen durch Niedersachsen, Holland und Engelland, Dritter Theil*, Rosten der Baumischen Handlung, Ulm.

Van den Spiegel, A. (1606), *Isagoges in rem herbariam libri duo*, Apud Paulum Meiettum, Patavii.

Vavilov, N.I. (1992), *Origin and Geography of Cultivated Plants*, Cambridge University Press, New York.

Vines, S.H. (1888), 'On the Relation between the Formation of Tubercles on the Roots of Leguminosae and the Presence of Nitrogen in the Soil', *Annals of Botany*, vol. 2, pp. 386–9.

Vines, S.H. (1896), 'Letter from the Sherardian Professor of Botany to the Chairman of the Committee. 12th June 1896', Reports relating to the Botanic Garden and to the Department of Botany 1875–1920, Sherardian Library of Plant Taxonomy, Bodleian Library.

Vines, S.H. (1911), 'Robert Morison (1620–1683) and John Ray (1627–1705)', in F.W. Oliver (ed.), *Makers of British Botany: A Collection of Biographies by Living Botanists*, University Press, Cambridge, pp. 8–43.

Vines, S.H., and G.C. Druce (1914), *An Account of the Morisonian Herbarium in the Possession of the University of Oxford*, Clarendon Press, Oxford.

Walters, S.M. (1981), *The Shaping of Cambridge Botany*, Cambridge University Press, Cambridge.

Ward, N.B. (1852), *On the Growth of Plants in Closely Glazed Cases*, John von Voorst, London

Waterston, C.D., and A. Macmillan Shearer (2006), *Former Fellows of the Royal Society of Edinburgh 1783–2002: Biographical Index*, part 2, Royal Society of Edinburgh, Edinburgh.

Watson, J.A.S. (1939), *The History of the Royal Agricultural Society of England, 1839–1939*, Royal Agricultural Society, London.

Watson, J.A.S., and P. Osborne (2004), 'Somerville, Sir William (1860–1932)', in *Oxford Dictionary of National Biography*, doi: 10.1093/ref:odnb/36193.

White, F. (1983), *The Vegetation of Africa: A Descriptive Memoir to Accompany the Unesco/AETFAT/UNSO Vegetation Map of Africa*, UNESCO, Paris.

Whitfield, J. (2012), 'Rare specimens', *Nature*, vol. 484, pp. 436–8.

Wilson, D. (2017), *Superstition and Science: Mystics, Sceptics, Truth-Seekers and Charlatans*, Robinson, London.

Wood, A. (1796), *The History and Antiquities of the University of Oxford*, vol. 2, printed for the editor, Oxford.

Wood, J.R.I., P. Muñoz-Rodríguez, B.R.M. Williams and R.W. Scotland (2020), 'A Foundation Monograph of *Ipomoea* (Convolvulaceae) in the New World', *PhytoKeys*, vol. 143, pp. 1–823.

Woodward, J. (1696), *Brief instructions for making observations in all parts of the world: as also for collecting, preserving, and sending over natural things being an attempt to settle an universal correspondence for the advancement of knowledge both natural and civil*, Richard Wilkin, London.

Worling, P.M. (2005), 'Pharmacy in the Early Modern World, 1617 to 1841 AD', in S. Anderson (ed.), *Making Medicines: A Brief History of Pharmacy and Pharmaceuticals*, Pharmaceutical Press, London, pp. 57–76.

Yanni, C. (2005), *Nature's Museums: Victorian Sciences and the Architecture of Display*, Princeton Architectural Press, New York.

Zirkle, C. (1935), *The Beginnings of Plant Hybridization*, University of Pennsylvania Press, Philadelphia.

Zirkle, C. (1951), 'Gregor Mendel and his Precursors', *Isis*, vol. 42, pp. 97–104.

INDEX

Numbers in **bold** indicate illustrations

First published in 2021 by the Bodleian Library,
Broad Street, Oxford OX1 3BG, in association with
the University of Oxford Botanic Garden and
Arboretum
www.bodleianshop.co.uk

ISBN: 978 1 85124 561 1

Publisher: Samuel Fanous
Managing Editor: Deborah Susman
Editor: Janet Phillips
Picture Editor: Leanda Shrimpton
Production Editor: Susie Foster
Designed and typeset by Dot Little in 10.5/15
   Freight Text
Printed and bound in Wales by Gomer Press
Limited on 150gsm Arctic matt paper

British Library Catalogue in Publishing Data
A CIP record of this publication is available from
the British Library